Contents

Preface

This book is intended to cover the various Medical (Health) Physics options to be examined at A-level and AS-level from 1996 onwards. Specifically, it is aimed at Module Ph7 of the NEAB syllabuses in A-level and AS-level Physics, Module PH3 (Topic 3B) of the London A-level Physics syllabus, Module 4835 of the Cambridge (modular) syllabuses in A-level Physics and A-level Science, Option M of the Cambridge (linear) syllabuses in A-level and AS-level Physics, Unit P4 of the Oxford and Cambridge syllabuses in A-level Physics and A-level Science, and Module 5 (Sections C and D) of the Oxford syllabuses in A-level and AS-level Physics.

The book contains many worked examples. Questions are included at relevant points in the text so that students can obtain an immediate test of their understanding of a topic. 'Consolidation' sections stress key points and in some cases present an overview of a topic in a manner which would not be possible in the main text. Definitions and fundamental points are highlighted – either by the use of screening or bold type. Questions, most of which are taken from past A-level papers, are included at the ends of the chapters.

Acknowledgements

I wish to thank Veronica Hilton for her assistance with proof-reading and with the preparation of the index. I also wish to express my gratitude to the publishers for their invaluable help throughout. Many other people have helped in a wide variety of ways – my sincere thanks to them and in particular to David Lloyd of the National Radiological Protection Board (NRPB), Dr Jane Lyons and Jean Murray of the Fairfield Hospital in Bury, and to Paul Maskens of the John Radcliffe Hospital (Oxford).

I am indebted to the following examination boards for permission to use questions from their past examination papers:

Associated Examining Board [AEB]
University of Cambridge Local Examinations Syndicate [C], reproduced by permission of University of Cambridge Local Examinations Syndicate
Cambridge Local Examinations Syndicate, Overseas Examinations [C(O)]
Northern Examinations and Assessment Board (formerly the Joint Matriculation Board) [N]
Oxford and Cambridge Schools Examinations Board [O & C]
Southern Universities' Joint Board [S]
University of London Examinations and Assessment Council (formerly the University of London School Examinations Board) [L]

R. MUNCASTER
Helmshore

A-Level

PHYSICS

Medical
Physics

Roger Muncaster

B.Sc Ph.D
Formerly Head of Physics
Bury Metropolitan College
of Further Education

Stanley Thornes (Publishers) Ltd.

First published in 1996 by
Stanley Thornes (Publishers) Ltd
Ellenborough House
Wellington Street
Cheltenham
Gloucestershire GL50 1YW
England

97 98 99 00 / 10 9 8 7 6 5 4 3 2

A catalogue record for this book is available from the British Library.

ISBN 0-7487-2324-2

The front cover shows MRI of brain.
The photograph is by courtesy of the Science Photo Library.

Photograph acknowledgements

Thanks are due to the following for providing photographs:
p55 Dr Ray Clark and Mervyn Goff/Science Photo Library
p61 Will and Deni McIntyre/Science Photo Library
p68 CNRI/Science Photo Library
p75 Simon Fraser/Dept. of Neuroradiology, Newcastle General Hospital/Science Photo Library
p77 Scott Camazine/Science Photo Library
p91 Susan Leavines/Science Photo Library
p95 GJLP/CNRI/Science Photo Library
p102 Science Photo Library
p113 Jean-Loup Charmet/Science Photo Library
p118 Martyn F. Chillmaid
p120 National Radiological Protection Board
p129 CNRI/Science Photo Library

Typeset by Tech-Set, Gateshead, Tyne & Wear.
Printed and bound at Scotprint, Musselburgh.

1
HUMAN MECHANICS

1.1 BONES AND JOINTS

The human skeleton consists of about 200 **bones**. Junctions between neighbouring bones are called **joints**. Although some joints (e.g. those in the skull) allow little or no movement, the majority enable the bones to move freely. These are known as **synovial joints,** and different types allow different types of movement. The elbow, for example, is a simple hinge, whereas the shoulder and the hip, which have a greater range of movement, are ball-and-socket joints.

The bones on either side of a synovial joint are held together by **ligaments**. The parts of the bones that form the joint are covered with **cartilage** – a tough, slightly elastic material that protects the bones from damage. Friction between the layers of cartilage is minimized by the presence of a lubricant known as **synovial fluid.**

1.2 MUSCLES

Bones are moved by **muscles** which are attached to them by **tendons**. When a muscle contracts (under the action of nerve impulses from the brain) it pulls on the bones on each side of it and causes one of them to move. Muscles cannot 'push', and therefore to return the bone to its original position, the first muscle relaxes and a second muscle, acting in opposition to the first, contracts (Fig. 1.1). The two muscles are known as an **antagonistic pair**.

Fig. 1.1
An antagonistic pair: the biceps and triceps

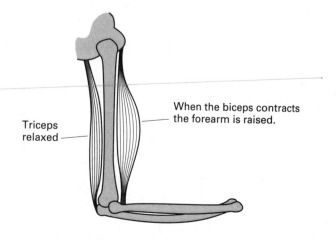

Triceps
relaxed

When the biceps contracts
the forearm is raised.

1.3 LEVERS

When a muscle contracts, it exerts a force on the bone at the point where it is attached to it by the tendon, producing a turning moment about the joint. The system acts as a lever with the joint as the **fulcrum**. The force exerted by the muscle is the **effort**, and the weight of the part of the body being moved (together with that of anything attached to it) is the **load**.

EXAMPLE 1.1

The forces acting on the forearm illustrated in Fig. 1.2 are its weight (20 N), the weight of the book (10 N), the effort exerted by the biceps muscle (E), and the reaction due to the upper arm (R). Calculate E and R.

Solution

Taking moments about F gives

$$E \times 4.0 = 20 \times 14 + 10 \times 32$$

$$\therefore \quad 4.0E = 280 + 320$$

$$\therefore \quad E = \frac{600}{4.0} = 150$$

i.e. Effort due to biceps $= 1.5 \times 10^2 \, \text{N}$

Resolving vertically gives

$$R + 20 + 10 = E$$

$$\therefore \quad R + 30 = 150$$

$$\therefore \quad R = 120$$

i.e. Reaction due to upper arm $= 1.2 \times 10^2 \, \text{N}$

Fig. 1.2
Diagram for Example 1.1
showing the forces acting
on the forearm

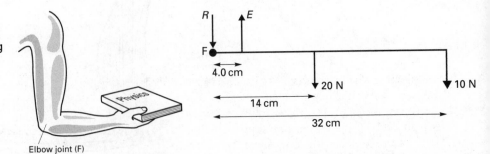

Notes (i) The lever in Fig. 1.2 is a **third-class** lever (i.e. effort in the middle). Examples of a **first-class** lever (fulcrum in middle) and a **second-class** lever (load in middle) are shown in Fig. 1.3.

Fig. 1.3
Forces acting on (a) the head (a first-class lever system) and (b) the foot (a second-class lever system)

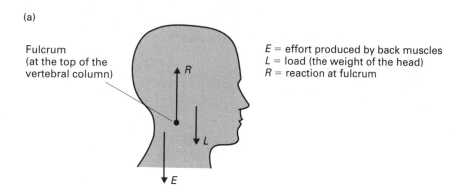

(a)

Fulcrum
(at the top of the
vertebral column)

E = effort produced by back muscles
L = load (the weight of the head)
R = reaction at fulcrum

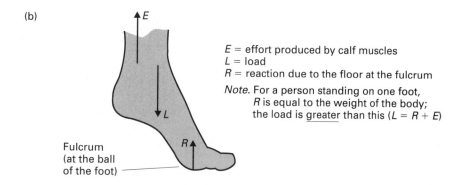

(b)

E = effort produced by calf muscles
L = load
R = reaction due to the floor at the fulcrum

Note. For a person standing on one foot, R is equal to the weight of the body; the load is <u>greater</u> than this ($L = R + E$)

Fulcrum
(at the ball
of the foot)

(ii) The **mechanical advantage** of a system is defined by

$$\text{Mechanical advantage} = \frac{\text{Load}}{\text{Effort}}$$

In the system shown in Example 1.1 the effort has to be bigger than the load because it acts closer to the fulcrum than the load does. Although this makes the mechanical advantage less than one, it means that **large movements of the arm are produced by only small contractions of the muscle.**

(iii) Most of the joints in the body are third-class levers. Since all third-class levers have a mechanical advantage which is less than one, it follows that **most of the joints are designed for speed of movement rather than for lifting heavy loads.**

QUESTIONS 1A

1. A weight-lifter whose mass is 80 kg holds a mass of 100 kg above his head whilst standing on tiptoe (on both feet). Assuming that the forces acting on each of his feet are as shown in Fig. 1.3(b), calculate the force exerted by each calf muscle, given that the horizontal distances between the effort and the fulcrum, and between the load and the fulcrum are respectively 10 cm and 4.0 cm. (Assume $g = 10\,\text{m}\,\text{s}^{-2}$.)

1.4 THE VERTEBRAL COLUMN (SPINE)

The spine consists of 33 **vertebrae** and provides the main support for the body. Nine of the vertebrae at the base of the spine are fused, five to form the **sacrum** and four to form the **coccyx** (Fig. 1.4). The top 24 are covered with cartilage and are separated from each other by tough fibrous pads known as **discs**. The discs allow the spine to bend and to twist. They also protect the vertebrae from wear and cushion them from shock.

Fig. 1.4
The vertebral column

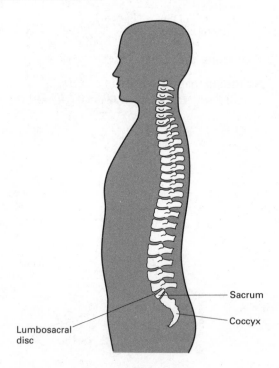

1.5 STANDING

The disc immediately above the sacrum, the **lumbosacral disc**, supports about 60% of the total body weight. In normal upright posture it lies at an angle of approximately 40° to the horizontal, and is therefore subject to both **compressive stress** (squashing) and **shear stress** (twisting). The situation is illustrated in Fig. 1.5 in which the disc is at an angle θ to the horizontal. There is a downward directed force of $0.6W$, where W is the total body weight, and this is balanced by an

Fig. 1.5
Forces acting on
lumbosacral disc

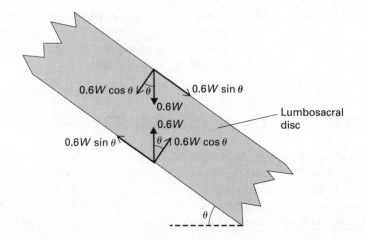

equal, upward directed force due to the reaction of the sacrum. Resolving parallel and perpendicular to the surface of the disc gives

$$\text{Shear force} = 0.6W \sin \theta$$

$$\text{Compressive force} = 0.6W \cos \theta$$

The structure of the disc is such that it is more likely to be damaged by shear than by compression. Pain in the lower back is often the result of adopting a posture in which the pelvis is tilted forward – this increases the value of θ and so increases the shear force.

Forces at the Hip

Assuming that the weight of each leg is $0.15W$ (where W is total body weight), the weight of the rest of the body is $0.7W$ and therefore when standing on two legs the force at the head of each femur (Fig. 1.6) is $0.35W$. Lifting one leg off the floor (which, incidentally, requires the body's centre of gravity to shift so that it is vertically above the other foot), creates a turning moment about the femoral head of the supporting leg, and the abductor muscle on the support side has to contract to counter it. The force at the head of the femur increases from $0.35W$ to about $2.5W$ – brought about in part by the extra weight it has to support but mainly by the contraction of the abductor muscle.

Fig. 1.6
Forces at the hip

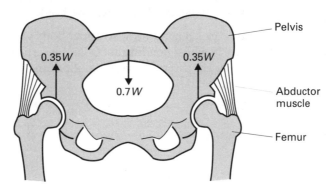

1.6 **BENDING AND LIFTING**

In Fig. 1.7 the trunk is bent forward at $70°$ to the vertical. The weight of the upper body ($0.6W$) acts through G and exerts a turning moment about the lumbosacral joint. This is countered by the pull (E) of the erector spinae muscles (which connect the spine to the pelvis) acting at about $10°$ to the spine, and which can be taken to act through G. The third force, R, is the reaction of the sacrum on the lumbosacral disc at the base of the spine. Since the three forces are in equilibrium, R must also act through G.*

Resolving perpendicular to the spine gives

$$E \sin 10° = 0.6W \sin 70°$$

$$\therefore \quad E = 3.25W$$

---

*It is a standard result that <u>three non-parallel coplanar forces</u> that are in equilibrium must all act through the same point – see, for example, R. Muncaster, *A-Level Physics* – section 4.2.

Fig. 1.7
The forces involved in
bending

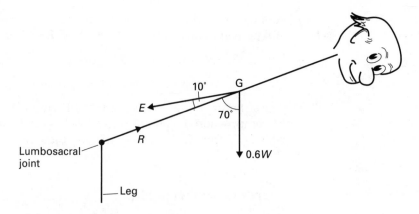

Resolving parallel to the spine gives

$$R = E \cos 10° + 0.6W \cos 70°$$

$$\therefore \quad R = 3.40W$$

Note that R, the force on the lumbosacral disc, is over three times body weight and nearly six times the force on it when standing upright. Lifting heavy weights with the back in this position puts even greater stress on the disc and might even cause it to rupture. **Damage is much less likely to occur if heavy objects are lifted with the knees bent and the back vertical**.

In Fig. 1.8 a man whose upper body weight is 500 N is lifting an object of weight 250 N whilst bending at 60° to the vertical. We no longer have a situation in which there are three forces in equilibrium and so R now acts at a small angle α to the spine.

Fig. 1.8
The forces involved in
bending and lifting

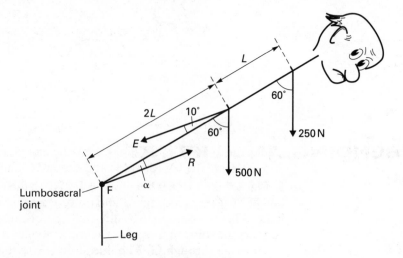

Taking moments about F gives

$$E \times 2L \sin 10° = 500 \times 2L \sin 60° + 250 \times 3L \sin 60°$$

$$\therefore \quad E = 4364 \ (= 4.36 \times 10^3 \, \text{N})$$

Resolving perpendicular to the spine gives

$$R \sin \alpha = E \sin 10° - 500 \sin 60° - 250 \sin 60°$$

$$\therefore \quad R \sin \alpha = 108.3$$

Resolving parallel to the spine gives

$$R\cos\alpha = E\cos 10° + 500\cos 60° + 250\cos 60°$$

$$\therefore \quad R\cos\alpha = 4673$$

Since

$$R^2 = (R\sin\alpha)^2 + (R\cos\alpha)^2$$

$$R^2 = 108.3^2 + 4673^2$$

i.e. $\quad R = 4.67 \times 10^3\,\text{N}$

1.7 WALKING AND RUNNING

When the heel strikes the ground at the end of a stride (Fig. 1.9(a)) the ground exerts an upward directed force, R, to destroy the leg's downward momentum, and a frictional force, F, to destroy its forward momentum. The resultant of these forces is called the **ground force** and it can be taken to act along the line of the leg. Its magnitude, G, is given by

$$G^2 = R^2 + F^2$$

Fig. 1.9
Ground force when walking (a) at heel-strike (b) at toe-off

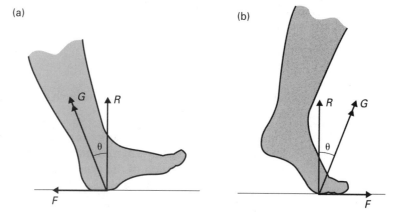

When the toe pushes off (Fig. 1.9(b)) the ground force provides the thrust required to accelerate the leg forwards and upwards. The horizontal force is provided by friction and therefore must not be required to exceed μR or else the foot will slip. From Fig. 1.9

$$\tan\theta = \frac{F}{R}$$

Therefore if no slipping is to occur

$$\tan\theta \leqslant \frac{\mu R}{R}$$

i.e. $\quad \tan\theta \leqslant \mu$

It follows that **slipping is more likely to occur**:

(i)   when μ is small (i.e. on slippery surfaces), and

(ii)   when θ is large (i.e. long strides).

We can run at up to six times faster than our normal walking speed. This is achieved by an increase in both stride length and stride frequency. The increased stride length is brought about by increasing the ground force so that we actually leap off the ground at each step. The body twists about its axis during running. As the right leg swings forwards, so too does the left arm, causing the upper and lower parts of the body to rotate in opposite senses and so conserve angular momentum.

In both walking and running, the centre of gravity of the body rises and falls during each step and therefore our leg muscles are required to do work against gravity. Energy is also used in providing the rotational kinetic energy of the legs as they swing back and forth at each stride.

1.8 ENERGY CONVERSIONS IN JUMPING AND LANDING

When we jump, our leg muscles do work to create the kinetic energy ($\frac{1}{2}mv^2$) required for take off. This is converted to gravitational potential energy (*mgh*) as the body's centre of gravity rises. It follows that

$$\text{Work done} = mgh$$

where h is the height through which the centre of gravity rises, and includes the height gained <u>before</u> take off when the jump starts from a crouching position.

On the basis of KE lost = PE gained, it is a simple matter to show that

$$h = \frac{v^2}{2g}$$

where v is the take off speed and h is the height through which the centre of gravity rises <u>after take off</u>.

Falling converts potential energy into kinetic energy, and this is destroyed on landing. Suppose a person of mass m hits the ground with speed v and is brought to rest in a time t as a result of the ground exerting an <u>average</u> force F. Since

$$\boxed{\text{Impulse} = \text{Change in momentum}}$$

$$Ft = mv$$

i.e. $$F = \frac{mv}{t}$$

It follows that the average force on the body, and therefore the risk of injury, can be reduced by prolonging the time taken <u>for the body as a whole</u> to come to rest – by flexing the knees, for example.

The ground deforms as a result of the impact, so too do the bones, cartilage and soft tissue. It is the kinetic energy that the body had immediately before the impact that has been used to produce these deformations. The severity of the impact determines which of these may be permanent, and to what extent.

Note When a person jumps, the height through which the <u>centre of gravity</u> rises after take off depends only on the speed of take off – it is not affected by any changes in posture once the body has left the ground. However, changes in posture do affect the height to which the head, for example, rises because they alter the position of the centre of gravity within the body (see Fig. 1.10).

Fig. 1.10
To show the effect of a
change of posture on the
height of a jump

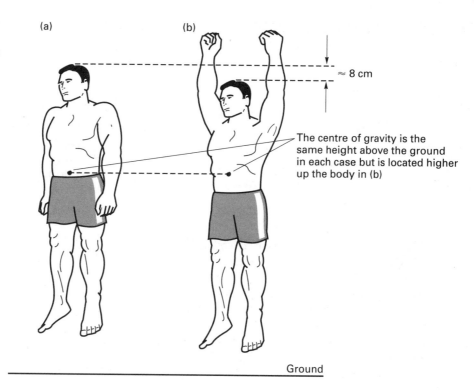

(a)   (b)

≈ 8 cm

The centre of gravity is the
same height above the ground
in each case but is located higher
up the body in (b)

Ground

QUESTIONS 1B

[Assume $g = 10 \, \text{m s}^{-2}$]

1. A girl of mass 40 kg drops vertically through a height of 60 cm. Assuming that her body is upright throughout, find the speed at which she hits the ground.

2. A man of mass 80 kg jumps vertically from a crouching position in which his centre of mass is 80 cm above the ground.
   (a) Calculate the work done by his muscles if his centre of gravity is 170 cm above the ground at the top of the jump.

   (b) At what speed does he take off if his centre of gravity is 110 cm above the ground at the time?

3. A man of mass 60 kg walks along a horizontal road at a steady speed of 4.6 km h⁻¹ by taking strides of length 96 cm. His centre of gravity rises (and falls) 9.0 cm at each step. Calculate the rate at which he is doing work against gravity.

4. At what rate is a man of mass 60 kg working when he runs up a flight of stairs of height 2.4 m in 3.0 s?

CONSOLIDATION

The **bones** on either side of a **synovial joint** are held together by **ligaments**.

Bones are moved by **muscles** which are attached to them by **tendons**.

Most of the joints in the body are **third-class** levers and therefore are designed for speed of movement rather than for lifting heavy loads.

Bending increases the force on the **lumbosacral disc**. Heavy objects should be lifted with the knees bent and the back vertical.

Slipping is more likely to occur when taking long strides on slippery surfaces.

QUESTIONS ON CHAPTER 1

Assume, where necessary, that $g = 10\,\mathrm{m\,s^{-2}}$.

1.

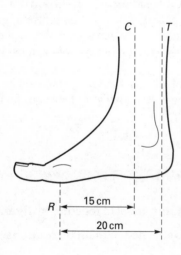

(a) The biceps muscle, shown as B in the diagram (not to scale), acts when a mass of 10 kg is held with the forearm horizontal so the angle at the elbow is 90°. Neglecting the mass of the forearm and hand, what is the force provided by the muscle when the hand is held at rest?

Is the energy consumed by the muscle equal to the work done on the mass? Give a reason for your answer.

(b) By what factor does the force provided by the biceps muscle increase when the arm is stretched out with the forearm still horizontal but with the angle $\theta = 30°$?

(c) By considering the results of your calculation, discuss how to support the mass in the hand so that the biceps muscle has to exert the least tension with the forearm still horizontal.   [N, '79]

2.

(a) The figure shows (in simplified form) the lines of action of the forces acting on the foot of a child. The child, whose weight is 400 N, is standing upright on two feet with heels just clear of the ground. The child's weight is distributed equally between the two feet.

R is the reaction (supporting force) from the ground.

C is the force exerted on the foot by the lower leg bones.

T is the force exerted on the foot by the Achilles tendon.

Calculate the magnitude and direction of R, C and T.

(b) The child now rises so that the heels are further from the ground. Discuss the effect of this on R, C and T.

Assume the lines of action of these forces are unchanged and that they remain the only forces acting.   [O & C, '92]

3. When lifting objects from the ground, it is preferable to bend the knees and keep the back as vertical as possible rather than keep the legs straight and bend forward from the hips. Explain this. Provide a diagram to support your explanation.   [N, '83]

4. (a) With the aid of simplified diagrams of the human body:

(i) Identify the principal forces involved when one 'touches one's toes' with legs and arms kept straight. Hence explain the consequences to the body of lifting heavy objects with a bent back.

(ii) Describe how the forces acting on the feet whilst walking differ from those acting whilst running.

(b) Estimate

(i) the compressional force in each arm when a gymnast of mass 60 kg does slow 'press-ups',

(ii) the mean power exerted when the press-up rate is increased to 1 per second.   [O & C, '91]

2

ENERGY EXPENDITURE

2.1 BASAL METABOLIC RATE

The body requires a regular intake of energy (from **carbohydrates**, **fats** and **proteins**) in order to perform its normal functions. The rate at which it uses this energy is called the **metabolic rate**. The body uses energy even when it is at rest, for it has to maintain brain activity, keep the heart beating and perform other involuntary activities such as breathing, cell replacement and the manufacture of various enzymes.

> The rate at which the body uses energy when it is completely at rest is called the **basal metabolic rate (BMR)**.

BMR varies with age as shown in Fig. 2.1. Children have high values because of the energy required for growing. On average, men have slightly higher values than women – primarily because men have less body fat and therefore use more energy in maintaining body temperature.

Fig. 2.1
Variation of basal metabolic rate with age

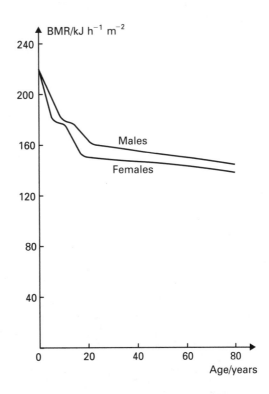

Note Surface area has a major influence on the rate at which the body consumes energy and therefore BMR values are usually expressed in kilojoules per hour <u>per square meter of body surface</u> ($kJ\,h^{-1}\,m^{-2}$).

QUESTIONS 2A

1. In a measurement of BMR a man is found to be consuming energy at a rate of 82.1 W. Charts show that a man of his height and weight has a surface area of $1.92\,m^2$. What is his BMR expressed in $kJ\,h^{-1}\,m^{-2}$?

2.2 DAILY ENERGY REQUIREMENT

We have to expend energy in order to maintain normal body temperature – both to keep warm when we are cold, and to keep cool when we are hot. Simple tasks such as walking, or even sitting at a desk, also consume energy. It follows that the daily energy requirement of an individual depends not only on their particular BMR (and therefore on their age and sex) but also on the ambient temperature, the amount of clothing worn, the surface area of the body, the amount of exercise undertaken, etc. The rates at which an average person consumes energy whilst performing various activities are shown in Table 2.1. The more vigorous the activity, the greater the consumption of energy in order to provide energy for the work done by the muscles and for the increased heart and breathing rates involved. **The <u>daily</u> energy requirement of an adult is usually in the range 1.0 to 1.6 MJ.**

The rate at which the body produces energy depends largely on tissue **volume**, whereas the rate at which energy is lost depends on body **surface area**. The smaller the body, the greater the ratio of surface area to volume. It follows that babies are more at risk of hypothermia than adults, and that small people need to eat more per kilogram of body mass than large people.

Table 2.1
Rate of energy consumption during various activities

Activity	Typical rate of consumption of energy/ W
Resting	80
Walking slowly	200
Walking quickly	350
Swimming	450
Playing football	600
Sprinting	1000

2.3 TEMPERATURE REGULATION

A complicated control mechanism maintains the various organs of the body at a fairly constant temperature of about 37 °C. This is known as the **core temperature** or **body temperature**. The temperature of the skin, on the other hand, is influenced by the ambient temperature and by a number of other factors. It can vary quite considerably but is typically about 4 °C below core temperature.

The body is extremely intolerant of changes in core temperature. The effects of **hypothermia** (lowering of body temperature) and **hyperthermia** (raising of body temperature) are summarized in Table 2.2. The heat control mechanism becomes less efficient with ageing, making elderly people susceptible to hypothermia when exposed to cold and damp. Hyperthermia can be induced by feverish illness or by over vigorous exercise.

Notes

(i)   A body temperature of over 39 °C is classed as **hyperthermia**; less than 35 °C is classed as **hypothermia**.

(ii)   An estimate of core temperature can be made simply by placing a mercury-in-glass **clinical thermometer** in the mouth and leaving it there for a couple of minutes to come into thermal equilibrium. A more accurate estimate is obtained by inserting a thermometer in the rectum to a depth of about 10 cm. Rectal thermometers are usually **thermistors**.

Table 2.2
The effects of changes in body (core) temperature

Temperature	Condition
Above 43 °C	Death.
41 °C	Damage to central nervous system. Convulsions.
39 °C	Dilation of peripheral blood vessels increases blood flow to skin and so increases rate of loss of heat. Increased heart rate. Reduced blood flow to brain and possible loss of consciousness.
37 °C	Normal.
35 °C	Shivering. Constriction of peripheral blood vessels reduces blood flow to skin and so reduces rate of loss of heat.
32 °C	Shivering ceases. Temperature regulation fails.
30 °C	Loss of consciousness.
Below 28 °C	Death.

The body controls its temperature by maintaining a balance between the rate at which it produces heat and the rate at which it loses it. It generates heat (primarily) by oxidizing food; it loses it (primarily) by conduction, convection, radiation, evaporation and respiration. (The body may also <u>gain</u> heat by the first three of these processes.) **The relative importance of the various heat loss mechanisms depends on the type of activity being undertaken, the environmental conditions and the extent to which the body is covered by clothing.**

Evaporation

Evaporation (sweating) is the dominant mechanism when the ambient temperature is high. As sweat evaporates from the skin it cools it by removing the necessary latent heat of vaporization. High humidity reduces the effectiveness of the process; air movement increases it. (Note. We lose heat by evaporation even when we are not visibly sweating – water is <u>always</u> evaporating from the skin unless there is 100% humidity.)

In extreme conditions, a person may lose as much as 1 litre of sweat per hour and consequently needs a large fluid intake to replace that which has been lost.

Convection

When cool air comes into contact with the body it draws heat from the body and so becomes warmer and less dense. It therefore rises and is replaced by more cold air which draws more heat away, and so on. The colder the air, the greater the cooling. Wind speed is also a factor, the higher the wind speed, the greater the effect – the so-called **wind chill factor**.

Conduction

This is the process by which heat is lost from the body through <u>contact</u> with colder objects such as a chair or the floor. Conduction normally accounts for no more than about 5% of the total heat loss, but there are circumstances (e.g. swimming in cold water) in which it accounts for very much more.

Radiation

Our bodies lose heat by radiation to <u>cooler</u> objects nearby. The greater the temperature difference, the greater the rate of loss of heat. We <u>absorb</u> heat from any object at a <u>higher</u> temperature than ourselves. **Note**. The rate at which we gain or lose heat through radiation is not affected by air temperature – a skier can remove his shirt in bright sunshine and remain comfortably warm even when the air temperature is very low.

Radiative heat loss is affected by posture – curling the body into a ball, for example, reduces the effective radiating area and so reduces heat loss.

Respiration

Exhaled air removes the heat it has gained (by conduction) through contact with the lungs and nasal passages. Heat is also lost as a result of moisture in the lungs evaporating and forming water vapour that is subsequently expelled along with the exhaled air. Heat loss through respiration is greatest when the air is both cold and dry.

The body has various means by which it maintains the balance between heat loss and heat production.

(i)   It may alter the **metabolic rate** (the rate at which it consumes energy).

(ii)   Blood loses heat as it flows through the capillaries near the surface of the skin. The capillaries widen (**vasodilation**) in warm conditions, thereby increasing the rate at which blood flows through them and so increasing the rate of loss of heat. The opposite effect (**vasoconstriction**) occurs in cold conditions.

(iii)   **Muscular activity generates heat**. (Most of the energy consumed by our muscles produces heat rather than useful work.) The body makes use of this when we are cold by causing rapid contractions of the muscles – **shivering**.

(iv)   The sweat glands become more active when we are hot.

There is a limit to what the body can do for itself and therefore we need to wear clothing suited to the environmental conditions and to whatever activity is being undertaken. We can also warm ourselves by exercising and by taking hot drinks.

In cold conditions, we need to wear clothes that trap air between the different layers of clothing. The volume of the trapped air is too small to allow convection currents to circulate and therefore since air is a poor <u>conductor</u> of heat, the trapped air insulates us from the surroundings. If our clothes are too tight, there is no trapped air and therefore little insulation. Clothes that are too loose do not keep us warm because the volume of air between adjacent layers is large enough for convection currents to circulate. The air pockets amongst the feathers of a duvet provide the same function as the air trapped between layers of clothing.

When clothes become wet they provide very little insulation because the trapped air is replaced by water which is a much better conductor of heat than air.

Heat lost through radiation can be reduced by wearing clothes that have a reflective coating on the inside. The survival suits worn by mountaineers employ this technique.

QUESTIONS 2B

1. Calculate the rate, in watts, at which a girl is losing heat by evaporation whilst sunbathing if she is producing sweat at a rate of $0.48\,\mathrm{kg\,h^{-1}}$. (The specific latent heat of vaporization of sweat at the temperature concerned is $2.4 \times 10^6\,\mathrm{J\,kg^{-1}}$.)

2.4 ESTIMATION OF THE POWER PROVIDED BY MUSCLES

This can be achieved by carrying out simple experiments designed to measure the power output of a specific set of muscles. The power of the leg muscles, for example, can be obtained by timing a person running up a flight of stairs or by determining the number of times he can raise his leg in a given time with a weight attached to his ankle.

EXAMPLE 2.1

A man of mass 75 kg runs up a flight of 50 steps in 15 seconds. Calculate the power output of his leg muscles given that the vertical height of each step is 0.20 m. (Assume $g = 10\,\mathrm{m\,s^{-2}}$.)

Solution

$$\text{Work done} = \text{Increase in gravitational PE}$$

$$= mgh$$

$$= 75 \times 10 \times (50 \times 0.20)$$

$$= 7500\,\mathrm{J}$$

$$\text{Power} = \frac{\text{Work done}}{\text{Time taken}}$$

$$= \frac{7500}{15}$$

$$= 5.0 \times 10^2\,\mathrm{W}$$

QUESTIONS ON CHAPTER 2

Assume, where necessary, that $g = 10\,\mathrm{m\,s^{-2}}$.

1. **(a) (i)** What do you understand by *basal metabolic* rate (BMR)?
   (ii) Why is the BMR of a child greater than that of an adult?

   (b) A flight of stairs consists of 60 steps, each of height 20 cm. A man of mass 80 kg claims to be able to run up the steps in 5.5 s. Make suitable calculations so that you can decide whether the claim is justified.

   (c) When climbing the steps in **(b)**, the body muscles are able to work with an efficiency of 17%.
   (i) Calculate the amount of energy wasted.
   (ii) Given that the specific latent heat of vaporisation of sweat is $2400\,\mathrm{kJ\,kg^{-1}}$, what mass of sweat must evaporate in order to dissipate this waste energy?

   (d) Discuss why, in practice, the mass of sweat excreted is likely to differ from that calculated in **(c)(ii)**.   [C, '94]

2. **(a)** What three types of food provide energy to a person?

   (b) The rate of energy supply needed by a person resting but awake is called the basal metabolic rate. (A typical value for a 20 year old woman is $1.1\,\mathrm{W\,kg^{-1}}$.) State three different needs which this rate of energy supply meets.

   (c) A woman whose mass is 55 kg has a metabolic rate of $9.0\,\mathrm{W\,kg^{-1}}$ when she is running up a hill at an angle of $5.0°$ to the horizontal with a constant speed of $6.0\,\mathrm{m\,s^{-1}}$.
   (i) At what rate is she gaining potential energy?
   (ii) At what rate is she using energy?
   (iii) Why are the answers to **(i)** and **(ii)** different?
   (iv) What is the effect on the woman of the fact that the answers to **(i)** and **(ii)** are different? What should she do to limit any adverse effect on her?
   (v) Suggest a reason why her kinetic energy is not constant when she is running up the hill at a constant speed.   [C, '91]

3. **(a) (i)** State what kind of thermometer you would use to measure the difference in temperature which occurs between inhaled and exhaled air during respiration. Give reasons for your choice.
   (ii) Explain briefly why body temperature rises during vigorous exercise, even though the heat lost from the body due to respiration increases.

   (b) (i) A person inhales in one breath $5.0 \times 10^{-4}\,\mathrm{m^3}$ of dry air at atmospheric pressure and $20\,°\mathrm{C}$. The air is then warmed to the body core temperature of $37\,°\mathrm{C}$ in the lungs. If the person takes 12 breaths per minute, calculate the heat transferred per minute to the air from the body. Assume that there are no pressure changes in the inhaled air during respiration.
   (ii) During each breath $2.2 \times 10^{-5}\,\mathrm{kg}$ of water vapour is formed in the lungs and is then expelled with the exhaled air. If the specific latent heat of evaporation of water at $37\,°\mathrm{C}$ is $2.3 \times 10^{6}\,\mathrm{J\,kg^{-1}}$, calculate the heat lost per minute due to this process. Comment on your result.

   Density of dry air at $20\,°\mathrm{C}$ and atmospheric pressure $= 1.2\,\mathrm{kg\,m^{-3}}$.

   Specific heat capacity of dry air at atmospheric pressure $= 1.0 \times 10^{3}\,\mathrm{J\,kg^{-1}\,K^{-1}}$.   [N, '91]

4. **(a)** Describe an experiment to estimate the average rate of working over a period of approximately five minutes which can be achieved by a student's leg muscles.

   (b) (i) On each beat of the heart, approximately 1 J of useful work is done, with an efficiency of about 15%. Estimate the power requirement of the heart for a resting person.
   (ii) The basal metabolic rate of an adult is approximately 85 W. Comment on the difference between this rate and your estimate obtained in **(b)(i)**.   [C, '92]

5. (a) Show that when a force F moves with velocity v in the direction of the force, the power supplied by the force is given by

$$\text{power} = Fv$$

(b) If a fluid is being pumped along a tube of cross-sectional area A with a constant velocity v, as shown in the diagram, the liquid flows from a region of high pressure p_1 to a region of low pressure p_2.

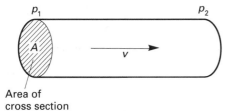

Area of
cross section

(i) What force is exerted on the fluid as a result of this pressure difference?

(ii) Deduce the power supplied to the fluid in terms of the pressure difference and the volume rate of flow of fluid.

(c) Calculate the power generated by the heart in order to maintain blood flow around the body of a healthy, resting person if the volume rate of flow of blood is $8.6 \times 10^{-5}\,\text{m}^3\,\text{s}^{-1}$ and the pressure difference from the arterial to venous systems is $12.8\,\text{kPa}$.

(d) The rate of energy usage of the heart referred to in **(c)** is likely to be several times the value you have just calculated. State two other uses for the input of energy to the heart in addition to the energy used in pumping blood.

(e) Suggest what is likely to happen to the two values quoted in **(c)** and to your answer to **(c)** for a person whose arteries are narrowed, possibly by cholesterol. [C, '93]

6. (a) The basal metabolic rate (BMR) varies from individual to individual but follows a general pattern with respect to gender and age.

(i) Explain the term *basal metabolic rate*.

(ii) Explain why metabolic rate increases during physical activity.

(b) An athlete expends $1.62\,\text{MJ}$ in total during a run lasting 30 minutes. The average efficiency of the athlete's muscles is 30%.

(i) Show that the average rate of production of thermal energy by the athlete is $630\,\text{W}$.

(ii) During the athlete's run, 40% of the thermal energy produced is transferred from the body through perspiration.
Calculate the mass of sweat which evaporates as a result of the run.
($2.3 \times 10^3\,\text{J}$ is required to evaporate $1.0\,\text{g}$ of sweat.)

(iii) State and explain other mechanisms by which excess thermal energy may be removed from the body.

(c) Suggest why an overweight person feels hot after only moderate exercise. [C, '95]

7. The basal metabolic rate (BMR) for a person depends upon the age and sex of that person.

(a) What is meant by *basal metabolic rate*?

(b) Use the data in the table to explain how and why the basal metabolic rate for an adult

(i) is different from that for a child,

(ii) varies with increasing age.

	Child	Young adult	Old adult
Average daily energy expenditure/MJ	5	12	8
Basal metabolic rate/kJ m^{-2} h^{-1}	200	160	140
Fat composition as a % of body weight	7	20	8
Surface area/m^2	0.9	1.8	1.4
Body volume/m^3	25×10^{-3}	71×10^{-3}	48×10^{-3}

(c) (i) Explain the importance of surface area and body volume in the energy requirement of a person.

(ii) Use the data in the table to calculate the ratio of surface area to body volume of an old adult and for a young adult.
Explain the significance of the result.

(d) Jane is a young adult who is fasting for a 24-hour period.

(i) Use the data in the table to show that she expends about $7\,\text{MJ}$ on the support of metabolic processes during this 24-hour period.

(ii) Hence find the work done by Jane on activities during this period of time.

(iii) Calculate the mass loss of body fat which Jane experiences during this time. Assume that all of the energy expended was at the expense of body fat. ($1.0\,\text{kg}$ of fat releases $38\,\text{MJ}$ of energy when 'burned'.) [C, '95]

3

VISION

The reader should be familiar with the properties of lenses, the lens formula and the concept of refractive index before continuing. We shall use the 'real is positive' sign convention.*

3.1 THE EYE AND ITS FUNCTION

An eye (Fig. 3.1) produces a real, inverted image of the object being viewed. The image is produced on the **retina** – the light-sensitive region at the back of the eye. The shape, and therefore the focal length, of the eye **lens** can be altered by the action of the **ciliary muscles** attached to it. This makes it possible for light from objects which are at different distances from the eye to be brought to a focus on the retina even though it is at a <u>fixed</u> distance from the lens. This ability of the eye is known as **accommodation**. (Compare this with the action of a camera, where objects at different distances are focused by altering the distance between the lens and the film.)

Fig. 3.1
Section through an eye

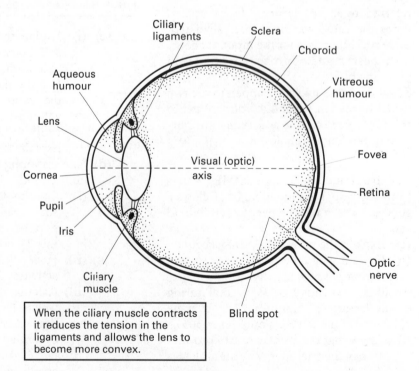

When the ciliary muscle contracts it reduces the tension in the ligaments and allows the lens to become more convex.

* See, for example, R. Muncaster, *A-level Physics* – sections 18.1–18.2 and 19.1–19.4.

The **aqueous humour** (a watery, transparent fluid) and the **vitreous humour** (a jelly-like, transparent fluid) provide nutrients and help to maintain the shape of the eyeball. The refractive index of the lens (Table 3.1) is not very different from that of the aqueous humour or that of the vitreous humour. Consequently light undergoes very little deviation as a result of passing through the lens itself. **The main function of the lens is to provide slight <u>changes</u> in deviation rather than large amounts of it. Most of the deviation occurs at the boundary between the air and the cornea** – where there is the largest change in refractive index.

Table 3.1
The refractive indices of the optical components of the eye

Medium	Refractive index
Air	1.00
Cornea	1.38
Aqueous humour	1.34
Surface of lens	1.39
Centre of lens	1.41
Vitreous humour	1.34

The iris is a (pigmented) diaphragm that controls the intensity of the light reaching the retina. It does this by adjusting the size of the **pupil** in response to signals from the retina – a negative feedback mechanism.

The sclera (the white of the eye) is the eye's tough outer cover. Its inner lining, the **choroid**, provides the blood supply for the retina. It contains a large amount of black pigment in order to reduce reflection of light within the eye and so prevent blurring of the image. The point at which the **optic nerve** joins the eye (from the brain) is known as the **blind spot** – so called because there are no light-sensitive receptors there.

3.2 NEAR POINT AND FAR POINT

The closest point on which an eye can focus (comfortably) is called its **near point**; the most distant point is called the **far point**. For a normal eye these are at 25 cm and infinity, respectively. Light from an object close to the eye has to be deviated more than that from a more distant object. This requires a more powerful, and therefore more highly curved, lens. It follows that the lens is at its most highly curved when focusing on the near point. An eye focused on infinity is relaxed and is said to be **unaccommodated**.

The near-point distance is known as the **least distance of distinct vision (D)**. An object whose distance from the eye is less than D appears blurred; one which is farther away appears smaller than when at the near point.

3.3 THE POWER OF A LENS

The power of a lens is defined by

$$\text{Power} = \frac{1}{\text{Focal length in metres}}$$ [3.1]

The unit of power is the **dioptre (D)**. Note. $1\,\text{D} = 1\,\text{m}^{-1}$.

The power of a converging lens is positive; that of a diverging lens is negative. It follows that a converging lens of focal length 25 cm has a power of 4 dioptres; a diverging lens of focal length 20 cm has a power of −5 dioptres.

Note An advantage of using power rather than focal length is that the overall power of a system of lenses (that are close together) is simply the algebraic sum of the component powers.

3.4 DEFECTS OF VISION

Short Sight (Myopia)

Nearby objects are seen clearly – distant ones are not, i.e. the far point is closer than infinity. Light from a distant object is brought to a focus in front of the retina (Fig. 3.2(a) and (b)). Either the eyeball is too long or the lens, at its weakest, is too strong. The condition is corrected by using a suitable diverging lens to make parallel rays of light appear to have come from the (uncorrected) far point of the eye (Fig. 3.2(c)). It should be clear from Fig. 3.2(c) that the diverging lens must have a focal length which is equal to (minus) the uncorrected far-point distance.

Fig. 3.2
(a) and (b) Short sight; (c) correction of short sight. (The eye lens is at its weakest in every case.)

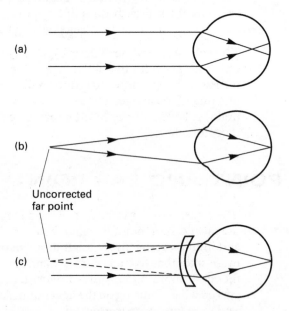

(a)

(b)

Uncorrected
far point

(c)

The use of a <u>diverging</u> lens makes the overall power of the eye/lens combination less than that of the eye alone; it therefore increases the near-point distance. (See Example 3.1.) This is not usually much of a problem – most short-sighted people have near points that are closer than normal anyway.

As an alternative to spectacles (or contact lenses) it is now possible to correct short sight by using a laser to reduce the curvature of the cornea, and so decrease the extent to which light is bent as it enters the eye.

Long Sight (Hypermetropia)

Distant objects are seen clearly – nearby ones are not, i.e. the near point is more than 25 cm from the eye. Light from a nearby object heads towards a focus behind the retina (Fig. 3.3(a) and (b)). Either the eyeball is too short or the lens, at its strongest, is too weak. The condition is corrected by using a suitable converging lens. The lens must produce a virtual image at the (uncorrected) near point of the eye of an object which is 25 cm from the eye (Fig. 3.3(c)).

Fig. 3.3
(a) and (b) Long sight; (c)
correction of long sight.
(The eye lens is at its
strongest in every case.)

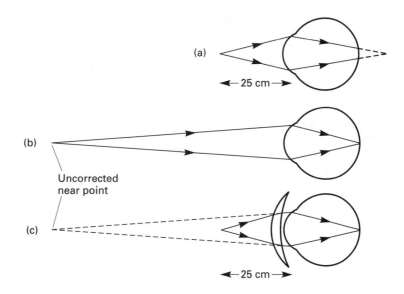

The focal length of the correcting lens is easily calculated. Suppose that a long-sighted person has a near-point distance of 150 cm. We see from Fig. 3.3 that an object 25 cm from the lens must produce a virtual image which is 150 cm from the lens, i.e. $u = 25$ cm and $v = -150$ cm. Therefore by $1/u + 1/v = 1/f$, $f = 30$ cm. Thus a converging lens with a focal length of 30 cm is required.

Note People who would normally need glasses are often able to read in bright sunlight. This is because the pupil constricts and so focuses the light in the same way as a pin-hole camera – there is no need to rely on accommodation. (The pin-hole restricts the light from each point on the object to a narrow cone.) A similar effect can be achieved in relatively dim light by holding a pin-hole close to the eye.

Astigmatism

This is usually due to the surface of the cornea not being spherical. The eye has different focal lengths in different planes, so that the vertical bar of a plus sign, for example, might be in focus when the horizontal bar is not. Astigmatism can be corrected by using a suitably oriented cylindrical lens. Fig. 3.4 illustrates a simple test – a person suffering from astigmatism will see some groups of lines in sharper focus than others.

Fig. 3.4
A simple test for
astigmatism

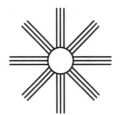

EXAMPLE 3.1

A man has a near-point distance of 20.0 cm and a far-point distance of 200 cm. What is his near-point distance when he is wearing spectacles that correct his far-point distance to infinity?

Solution

Since the man's far-point distance is 200 cm, his spectacles must have diverging lenses of focal length 200 cm (so that light from infinity appears to have come from 200 cm away).

Fig. 3.5
Diagram for Example 3.1

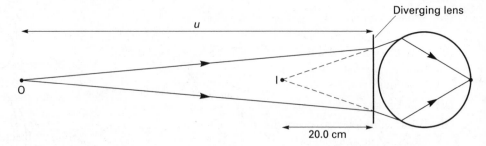

The closest point on which he can focus whilst wearing the spectacles is such that the light from it appears to have come from a point 20.0 cm from the eye after it has been refracted by the spectacle lens (see Fig. 3.5). Thus we need to find the object distance, u, for which the spectacle lens produces a virtual image distance of 20.0 cm. By

$$\frac{1}{u} + \frac{1}{v} = \frac{1}{f}$$

$$\frac{1}{u} + \frac{1}{-20.0} = \frac{1}{-200}$$

$$\therefore \quad \frac{1}{u} = -\frac{1}{200} + \frac{1}{20.0} = \frac{-1+10}{200} = \frac{9}{200}$$

$$\therefore \quad u = \frac{200}{9} = 22.2 \, \text{cm}$$

i.e. the near-point distance with the spectacles is 22.2 cm.

EXAMPLE 3.2

An eye focused on infinity has a power of 59 D. What is the power of the same eye when focused on a point 25 cm away?

Solution

Combining equation [3.1] and the lens formula $(1/u + 1/v = 1/f)$ gives

$$P = \frac{1}{u} + \frac{1}{v} \text{ (where } u \text{ and } v \text{ are in metres)}$$

We can regard the optical system of the eye as a simple lens at a distance d from the retina. Therefore

$$59 = \frac{1}{\infty} + \frac{1}{d} \quad \text{i.e. } \frac{1}{d} = 59$$

and

$$P_{25} = \frac{1}{0.25} + \frac{1}{d} \quad \text{i.e. } P_{25} = 4 + 59 = 63 \text{ dioptres}$$

where P_{25} is the power of the eye when it is focused 25 cm (i.e. 0.25 m) away.

QUESTIONS 3A

1. A girl can see clearly only those objects which are more than 250 cm from her eyes.

    (a) What defect of vision does she have?

    (b) What type of lens (converging or diverging) would be used to correct the defect?

    (c) What would be the focal length of the lens that would allow her to see objects 25 cm from her eyes?

    (d) What is the power of this lens?

    (e) What would her far-point distance be when wearing the spectacles?

2. An eye focused on a point 20 cm away has a power of 55 D. What is the power of the same eye when focused 200 cm away?

3. A man wears spectacles whose lenses have a power of +2.5 D in order to correct his near point to 25 cm. What is his near point distance when he is not wearing the spectacles?

4. A boy has a near point of 40 cm and a far point of 400 cm.
    (a) What spectacles are required to make his far point infinity?
    (b) What is his range of vision when wearing the spectacles?

5. A girl can see objects clearly only if they are between 60 cm and 600 cm away.
    (a) What spectacles are required to allow her to see objects that are only 25 cm away?
    (b) What is the most distant point on which she can focus whilst wearing the spectacles?

3.5 DEPTH OF FIELD AND DEPTH OF FOCUS

An eye cannot simultaneously focus (exactly) on objects that are at different distances from it. In Fig. 3.6(a) light from a point object at O is brought to a focus at I on the retina. Light from objects at O_1 and O_2 is out of focus and, in each case, covers an area whose diameter is XY and which is known as a **circle of confusion**. The range of object distances for which the circles of confusion are so small that the image is acceptably sharp is called the **depth of field**. Thus if XY is the diameter of the largest acceptable circle of confusion, the depth of field is O_1O_2.

Fig. 3.6
(a) To illustrate depth of field (b) To show the effect of reducing the diameter of the pupil

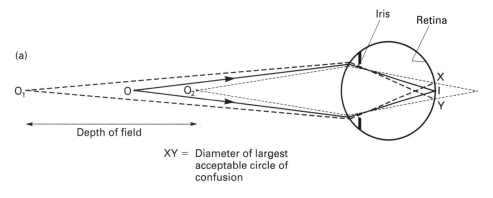

XY = Diameter of largest acceptable circle of confusion

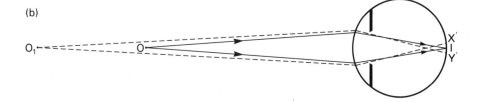

Decreasing the size of the pupil increases the depth of field because it decreases the diameter of the circle of confusion – X'Y' in Fig. 3.6(b) is smaller than XY in Fig. 3.6(a).

The depth of field is much greater for distant objects than for nearby objects. This is because any given change in object distance has a greater effect on image distance when the object distance is small than it does when it is large. (The reader is invited to confirm this by drawing simple ray diagrams or by putting numbers in the lens formula. This also explains why O is closer to O_2 than to O_1 in Fig. 3.6(a).) Our pupils constrict when we observe objects that are close by in order to offset (to some extent) the decreased depth of field.

The range of image distances over which the image of an improperly focused object is acceptably sharp is called the **depth of focus** (Fig. 3.7). It should be clear from Fig. 3.7 that **decreasing the size of the pupil increases the depth of focus**.

Fig. 3.7
To illustrate depth of focus

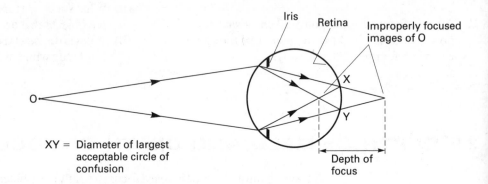

XY = Diameter of largest
acceptable circle of
confusion

3.6 THE RETINA

The retina (Fig. 3.8) contains light-sensitive cells known as **rods** and **cones**. There are about 120 million rods and 6 million cones. They are connected to nerve fibres which pass over the surface of the retina before joining together to form the **optic nerve**. Light has to pass through this layer of nerves before it can stimulate the rods and cones!

Fig. 3.8
The retina

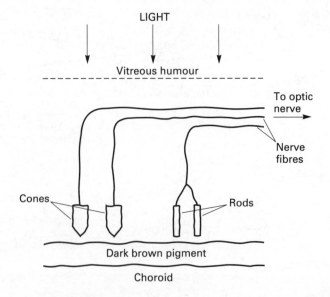

Rods cannot distinguish different colours but they allow us to see at much lower light intensities than cones because they have greater sensitivity to light and because several of them are connected to a single nerve fibre. (The fibre is triggered by the cumulative effect of stimuli which individually would be too weak.) Because they are grouped in this way, they give little perception of detail.

Cones allow us to distinguish different colours because there are three types of them, each sensitive to a different colour. They share fewer nerve fibres than rods and therefore allow us to see fine detail. Cones are effective only in bright light – which is why we cannot see colours when the light is dim.

Rods and cones are not evenly distributed over the retina. The **fovea** (Fig. 3.1) contains only cones, but the proportion of rods increases with distance from the fovea so that the edge of the retina consists almost entirely of rods. The cones in the fovea are each connected to a single nerve fibre. Because of this and because there are no nerve fibres covering it (they are arranged radially around it), **the fovea is the region of most acute vision**.

3.7 ADAPTATION OF THE RETINA

The pupil can vary in diameter from about 1.5 mm to about 8 mm, corresponding to a change in area, and therefore in the amount of light entering the eye, of a factor of about 30. Change in pupil size is primarily a response to <u>sudden</u> changes in illumination, it cannot account for the fact that we can see relatively well in moonlight and in sunlight, which can be over 10^5 times brighter. The ability of our eyes to adapt in this way is due mainly to a variation in the sensitivity of the retina itself. There are two processes – dark adaptation and light adaptation.

Dark Adaptation

Dark adaptation is the process by which the sensitivity of the retina increases when the light intensity decreases.

Rods and cones contain pigments which decompose on exposure to light, stimulating the associated nerve cells in the process. Enzyme activity subsequently regenerates the pigments – slowly in the rods, more quickly in the cones. In very bright light the pigment in the rods is broken down so much more quickly than it can be regenerated that only the cones remain active. If the level of illumination suddenly falls, vision is difficult at first because the rods are inactive. (It is particularly difficult if the intensity is so low that the cones are also inactive!) However, because the rate of decomposition is now much lower, the pigment concentration gradually builds up and so increases the sensitivity of the retina. Dark adaptation is more or less complete after about 30 minutes.

Light Adaptation

Light adaptation is the process by which the sensitivity of the retina decreases when the light intensity increases.

Sudden exposure to a higher level of illumination causes the photosensitive pigments in the rods and cones to decompose at an increased rate. Since regeneration is a slow process (particularly in rods), it cannot keep up with the rate of decomposition and therefore the number of active rods and cones falls and the sensitivity of the retina decreases. It is a very much quicker process than dark adaptation – the eyes take only a few minutes to adapt to bright sunlight, for example.

3.8 PERSISTENCE OF VISION

When a visual stimulus is suddenly removed, the sensation it has produced on the retina takes a finite time to disappear. The time for which this **after-image** persists depends on the intensity of the stimulus. Flashes of bright light appear to be continuous at frequencies above about 50 Hz; flashes of low-intensity light fuse together at much lower rates – about 5 Hz.

Were it not for this persistence of vision, we would be aware of each of the individual frames that make up the 'moving' images we see on TV (60 frames per second) and on a cinema screen (24 frames per second).

3.9 SPATIAL RESOLUTION

The resolving power of the eye is a measure of its ability to resolve (i.e. discern) fine detail. It is defined as the minimum angular separation θ for which two points can be seen as separate. (**Note**. The smaller the value of θ, the greater the resolving power, i.e. the greater the ability to resolve detail.)

If two points of light are to be seen as separate (i.e. resolved), they must each stimulate a different receptor on the retina and there must be at least one unstimulated receptor between these. When the image is formed on the fovea, which is where the receptors (cones) are most closely packed and where each has its own nerve fibre, this gives a minimum separation of 3 μm and corresponds to an angular separation of about 1.5×10^{-4} radians. The resolution is much poorer away from the fovea because the receptors are farther apart and share nerve fibres.

Whenever light passes through an aperture, it is diffracted to some extent. The smaller the aperture, the greater the diffraction and the greater the likelihood that the images of two objects will overlap and therefore be indistinguishable, i.e. unresolved. According to an arbitrary criterion proposed by Rayleigh (**Rayleigh's criterion**) two point objects will be resolved if their angular separation θ is given by

$$\sin \theta = \frac{1.22\,\lambda}{d}$$

where λ is the wavelength of the light concerned and d is the diameter of the aperture. Putting $\lambda = 5.5 \times 10^{-7}$ m (peak response on the fovea) and $d = 1.5 \times 10^{-3}$ m (minimum pupil diameter) gives $\theta \sim 10^{-4}$ radians which is consistent with the value based on the structure of the retina.

3.10 SPECTRAL RESPONSE

The eye is able to distinguish different colours because it has three types of cone, each containing a different pigment and therefore each responsive to a different colour – one primarily to red, one to green and one to blue (Fig. 3.9). When light of any particular wavelength falls on the retina, the extent to which it stimulates the various cones determines the colour perceived by the brain. (Note that the overall sensitivity of the cones (see Fig. 3.9) has its maximum value at about 555 nm.)

Fig. 3.9
Relative sensitivity of the three types of cone

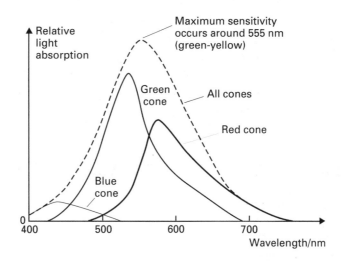

Although the rods provide no information about colour, they are not equally sensitive to all wavelengths, and have a peak response in the blue–green region around 510 nm.

QUESTIONS 3B

1. The wavelength at which the eye has its peak response depends on the intensity of illumination. State whether a decrease in light intensity shifts the peak response to longer or to shorter wavelengths. Explain your answer.

CONSOLIDATION

$$\text{Power of a lens} = \frac{1}{\text{Focal length in metres}} \qquad (\text{Unit} = \text{dioptre})$$

The near point is the position at which an object (or image) can be seen most clearly. An object (or image) closer than the near point appears blurred, one which is further away appears smaller than when at the near point.

The far point of an eye is the most distant point on which it can focus.

The eye lens provides slight changes in deviation rather than large amounts of it. Most of the deviation occurs at the boundary between the air and the cornea.

Short sight – nearby objects can be seen clearly, distant ones cannot. Corrected by using a diverging lens.

Long sight – distant objects can be seen clearly, nearby ones cannot. Corrected by using a converging lens.

Astigmatism – the eye has different focal lengths in different planes. Corrected by using a cylindrical lens.

Depth of field – the range of <u>object</u> distances for which the image is acceptably sharp. Increased by decreasing the size of the pupil.

Depth of focus – the range of <u>image</u> distances over which the image of an object is acceptably sharp. Increased by decreasing the size of the pupil.

Cones allow us to distinguish different colours and see fine detail but are effective only in bright light. There are three types (red, blue and green).

Rods cannot distinguish different colours and give little perception of detail but allow us to see in dim light.

QUESTIONS ON CHAPTER 3

1. **(a)** Using a labelled ray diagram in each case
    (i) explain what is meant by short sight,
    (ii) show how it is corrected.
    (b) A man can see clearly only objects which lie between 0.50 m and 0.18 m from his eye.
    (i) What is the power of the lens which when placed close to the eye would enable him to see distant objects clearly?
    (ii) Calculate his least distance of distinct vision when using this lens.   [N, '91]

2. **(a)** Explain with the aid of a ray diagram the action of a spectacle lens in correcting the far point of an eye from 200 cm in front of the eye to infinity. Repeat the procedure for correcting the near point of an eye from 50 cm to 25 cm.
    (b) In each case calculate the focal length of the spectacle lens, ignoring the separation of lens and eye.   [N, '78]

3. The combined power of cornea and lens of a normal, unaccommodated eye is 59 dioptres. If the eye focuses on an object 250 mm away find the change in power, in dioptres, of the eye.

    A myopic eye has a far point of 1.0 m and a near point of 150 mm. State the type of spectacle lens, needed for viewing an object at the normal far point and determine its power, in dioptres.

    Find the near point when wearing this lens.
    [N, '88]

4. A student complains that he is not able to see clearly any object unless it is more than 75 cm from his eyes. The normal near point is taken as being 25 cm from the eye.

    (a) **(i)** Name the student's eye defect.
    (ii) State what is meant by the *near point* of the eye.
    (b) **(i)** Copy the diagram on to your answer sheet.

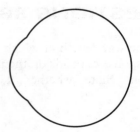

    On this, draw a ray diagram to illustrate the paths of two rays of light from a point object at the normal near point, showing how they would reach the retina of the student's eye.

    (ii) Draw a second ray diagram to show how a lens may be used to correct the defect for an object set at 25 cm from the eye.

    (iii) Calculate the power of this correcting lens.

(c) Some animals are able to change the curvature of the cornea so that they are able to see clearly both in air and in water. Explain why a change in curvature is necessary. [C, '94]

5. (a) Draw a diagram to illustrate long sight (hypermetropia) and a second diagram showing its correction by means of a suitable lens.

   Without spectacles an elderly person has a normal far point but is unable to see clearly objects closer than 200 cm.

   (i) Find the power in dioptres of the spectacle lens which alters the near point to 25 cm.

   (ii) When the person is wearing the spectacles find the far point and calculate the range of distinct vision.

   (b) Explain with the aid of a diagram the term *depth of focus* applied to the human eye.

   An object at a fixed distance from the eye is viewed when the pupil is (i) large and (ii) small. State and explain which condition gives rise to the larger depth of focus, supporting your answer by means of a diagram.

   (c) A person with normal eyesight observes a bright light which flashes for 0.020 s at intervals of 10 s. Explain how the person's perception of the bright light changes as the intervals are progressively reduced and name the effect which is responsible. State **one** practical application of the effect.
   [N, '90]

6. (a) Explain what is meant by the terms *short sight* and *astigmatism*. How may each of them be corrected by using spectacle lenses?

   (b) A man finds that he can focus clearly only on objects between 25 cm from the eye and infinity if he wears spectacles with lenses of power −2.5 dioptres.

   (i) What type of lens is this?

   (ii) What is the focal length of one of the lenses?

   (iii) What range of vision does he have if he is not wearing his spectacles?
   [C, '91]

7. (a) Draw a diagram of the human eye labelling the following features: *aqueous humour, cornea, iris, lens, retina* and *vitreous humour*. The diameter of the eyeball in your diagram should be about 10 cm.

   On your diagram draw a line (the optic axis) passing through the centre of the cornea and the centre of the lens and on to the retina. A ray of light travelling parallel to the optic axis enters the eye and eventually intersects the optic axis at the retina. Draw the path taken by this ray through the refracting system of the eye. Mark the place, with the letter P, where the major part of the refraction takes place.

   (b) The results of an eye test show that both eyes of the patient have near points at 0.20 m with far points located 0.50 m in front of the corneas.

   (i) Give the name of the vision defect indicated by these results.

   (ii) State the type of lens which would be used to correct it.

   (iii) Calculate the power, in dioptres, of the lens required to correct the defect.

   (iv) Calculate the new position of the near point when the patient is wearing the lens.

   (c) The left eye also shows some astigmatism.

   (i) Explain the nature of this defect of the refracting system.

   (ii) State the type of lens which would correct this condition. [N, '92]

8. (a) Compare what is seen by an observer, with normal eyesight, when a coloured object is illuminated by white light of (i) high intensity and (ii) low intensity. Give an explanation for your answer in terms of the behaviour of the eye.

   Sketch a graph showing the spectral response of the eye.

   (b) Two neighbouring, independent point sources of light are just resolved visually under optimum viewing conditions when they subtend 0.3 milliradians at the eye. If the sources are 0.5 m from the eye calculate their separation. Explain, in terms of the structure of the retina, how the two retinal images must be positioned to be seen as separate. [N, '86]

4

HEARING

4.1 THE NATURE OF SOUND

Sound is a longitudinal wave motion propagated by means of oscillations of the particles of the medium through which it is travelling. The human ear is sensitive to those sound waves with frequencies between about 20 Hz and 20 kHz. Frequencies above and below the range of human audibility are known as **ultrasonic** and **infrasonic**, respectively.

The speed c of sound in a fluid (i.e. a gas or a liquid) of density ρ and bulk modulus of elasticity K is given by

$$c = \sqrt{\frac{K}{\rho}} \quad \text{(for a fluid)} \qquad\qquad [4.1]$$

In a solid of density ρ and Young's modulus, E

$$c = \sqrt{\frac{E}{\rho}} \quad \text{(for a solid)} \qquad\qquad [4.2]$$

Notes (i) The speed of sound in a gas does not depend on the pressure of the gas. This follows from equation [4.1] (because both K and ρ are directly proportional to pressure) and is also confirmed by experiment.

(ii) It can be shown that for moderate pressures $c \propto T^{1/2}$, where T is the kelvin temperature of the gas.

(iii) The speed, c, frequency, f, and wavelength, λ, are related by

$$c = f\lambda$$

4.2 INTENSITY AND INTENSITY LEVEL: THE DECIBEL

The **intensity** of a sound wave is defined as the rate of flow of energy per unit area of a surface perpendicular to the direction of travel of the wave.

Note The intensity is proportional to the square of the amplitude of the wave, i.e.

$$\text{Intensity} \propto (\text{Amplitude})^2$$

The minimum intensity of sound that the human ear can detect is known as the **threshold of hearing** and is normally taken to be $1.0 \times 10^{-12}\,\mathrm{W\,m^{-2}}$ ($1.0\,\mathrm{pW\,m^{-2}}$). The maximum that can be experienced without the risk of permanent damage is about $100\,\mathrm{W\,m^{-2}}$. This is an enormous range, but it is not perceived as such by the ear – equal changes in intensity do not produce equal changes in **loudness**. Experiments indicate that **the ear has a logarithmic response to intensity**. Though it may not be obvious, this is equivalent to saying that increase in loudness is proportional to the <u>fractional</u> increase in intensity, i.e.

$$\text{Increase in loudness} \propto \frac{\text{Increase in intensity}}{\text{Initial intensity}} \qquad [4.3]$$

It can be shown that the loudness L of a sound of intensity I is given by

$$L = k\log_{10}\left(\frac{I}{I_0}\right)$$

where I_0 is the threshold intensity ($1.0 \times 10^{-12}\,\mathrm{W\,m^{-2}}$). It follows that the loudness increases by an amount k whenever the intensity increases by a factor of 10.

Note Equation [4.3] is an example of the **Weber–Fechner law** which states that sensations such as loudness and brightness increase in proportion to the fractional increase in the stimuli producing them.

Since the ear has a logarithmic response to intensity, it is useful to define a quantity that reflects this. Thus the **intensity level** or **relative intensity** of a sound of intensity I is defined by

$$\text{Intensity level in decibels} = 10\log_{10}\left(\frac{I}{I_0}\right) \qquad [4.4]$$

where I_0 is the **reference intensity**, normally taken to be $1.0 \times 10^{-12}\,\mathrm{W\,m^{-2}}$.

Whenever the intensity increases by a factor of 10, the intensity level increases by 10 decibels (10 dB) <u>regardless of the initial intensity</u> – see Table 4.1.

Table 4.1
Representative sound levels

Sound	Intensity level/dB	Intensity/ W m^{-2}
Threshold of hearing	0	10^{-12}
Rustling leaves	10	10^{-11}
Whispering	20	10^{-10}
Normal conversation	60	10^{-6}
Busy street	70	10^{-5}
Pneumatic drill	90	10^{-3}
Jet overhead	100	10^{-2}
Threshold of feeling	120	$10^{0} = 1$
Threshold of pain	140	$10^{2} = 100$

It is a simple matter to show this. It follows from equation [4.4] that if the intensity increases from I_1 to I_2 , then

$$\text{Increase in intensity level} = \left[10\log_{10}\left(\frac{I_2}{I_0}\right) - 10\log_{10}\left(\frac{I_1}{I_0}\right) \right] \text{dB}$$

$$= 10\left[\log_{10}\left(\frac{I_2}{I_0}\right) - \log_{10}\left(\frac{I_1}{I_0}\right) \right] \text{dB}$$

$$= 10\log_{10}\left(\frac{I_2}{I_1}\right) \text{dB}$$

$$= 10\,\text{dB} \quad \text{when} \frac{I_2}{I_1} = 10$$

Notes (i) A decibel (dB) is one tenth of a **bel (B)**, a unit named in honour of Alexander Graham Bell. The bel is too large for most purposes – the ear can detect changes of as little as about 1 decibel, hence its use.

(ii) An increase in intensity level of 1 dB corresponds to an increase in intensity of approximately 26% ($10\log_{10} 1.26 \approx 1$).

(iii) Intensity levels cannot be added and subtracted directly, they must first be converted to intensities.

QUESTIONS 4A

Assume, where necessary, that the reference intensity (I_0) is $1.0 \times 10^{-12}\,\text{W m}^{-2}$.

1. What are the intensity levels of sounds whose intensities are:
   (a) $1.0 \times 10^{-4}\,\text{W m}^{-2}$,
   (b) $3.0 \times 10^{-10}\,\text{W m}^{-2}$, (c) $7.0\,\text{pW m}^{-2}$?

2. What are the intensities of sounds whose intensity levels are:
   (a) $110\,\text{dB}$, (b) 68 dB? (Note, if $x = \log_{10} y$, then $y = 10^x$ – use the 10^x key on your calculator.)

3. A source of sound increases in intensity from $10^{-9}\,\text{W m}^{-2}$ to $10^{-7}\,\text{W m}^{-2}$. By how much does the intensity level increase?

4. The intensity level of the noise on a busy street is $70\,\text{dB}$. An aeroplane passes overhead and contributes a further $80\,\text{dB}$. What is the new intensity level?

5. The intensity level at a distance of $4.0\,\text{m}$ from a point source of sound is $40\,\text{dB}$. What is the power output of the source?

4.3 FREQUENCY RESPONSE

The **range of frequencies** which the human ear can detect varies considerably from one individual to another. The lower limit is typically around 20 Hz. The upper limit decreases with age but is about 20 kHz for an average young adult. The threshold of hearing depends very much on the frequency at which it is measured (see Fig. 4.1) and has its minimum value at about 3 kHz. Note that the **threshold of feeling**, the level at which the sensation changes from that of hearing to one of discomfort, is close to 120 dB at all frequencies.

Fig. 4.1
Frequency range of a
normal ear

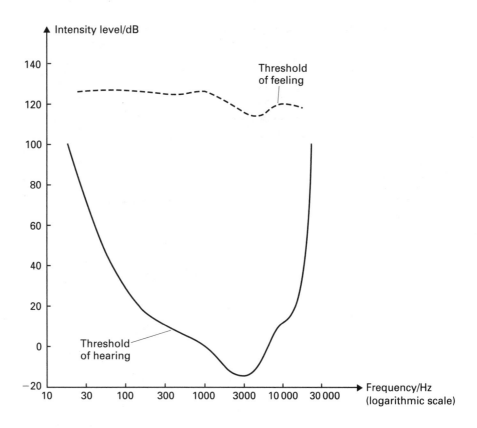

Note The decibel is defined in such a way that the intensity level of the threshold of hearing is 0 dB at 1 kHz. The threshold levels around 3 kHz, where the ear is more sensitive, are therefore negative.

Frequency discrimination is the ear's ability to distinguish different frequencies and is most acute at low frequencies. Between 60 Hz and 1 kHz frequencies as close as 3 Hz can be distinguished. This ability decreases with increasing frequency and is practically non-existent above 10 kHz.

4.4 LOUDNESS

Loudness is a subjective quantity – it depends on the intensity of the sound <u>and</u> on the hearing of the listener. A sound which is regarded as being loud by one person appears less loud to a person whose hearing is poorer. Furthermore, because the ear has different sensitivities to different frequencies, two notes which are of equal intensity, but which differ in pitch, may not sound equally loud, even to a single observer. (We cannot hear ultrasonic sounds, no matter how intense they are.)

The reader must not think that intensity level is a measure of loudness. Although it takes account of the logarithmic response of the ear to intensity, it takes no account of individual observers nor of the frequency response of the ear. It is also worth noting that **the decibel is a unit of intensity level; it is not a unit of loudness**.

The phon is a unit of loudness; it takes account of the fact that loudness is <u>frequency</u> dependent. To measure the loudness of a sound in phons, the source of the sound is placed next to a standard source with a frequency of 1000 Hz. The intensity of the standard source is then adjusted until the two sources are judged, by a normal observer, to be equally loud. If the standard source then has an intensity level of n decibels, the loudness of the sound being measured is n phons.

Measurements obtained in this way can be used to produce **curves of equal loudness** (Fig. 4.2). The curves illustrate the fact that loudness becomes less frequency dependent as the level of loudness increases. Note too that loudness and intensity level are numerically equal at 1 kHz.

Fig. 4.2
Curves of equal loudness for a normal ear

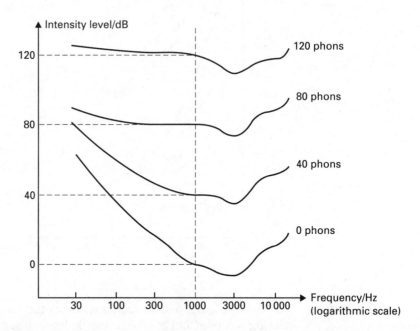

Another way of taking account of the frequency dependence of loudness is to use a sound level meter calibrated on a scale known as the **dBA scale**. The meter is essentially a microphone coupled to electronic circuitry that suppresses certain frequencies in such a way that the response of the microphone mimics that of a normal ear.

Sounds which have the same loudness (either in phons or dBA), but which have different frequencies, will not necessarily sound equally loud to someone with defective hearing – the two scales are based on the frequency response of a normal ear.

4.5 SENSITIVITY

The sensitivity of the ear is a measure of its ability to detect small changes in intensity. It is strongly frequency dependent and is defined by

$$\text{Sensitivity} = \log_{10}\left(\frac{I}{\Delta I}\right)$$

where $\Delta I/I$ is the smallest fractional change in intensity the ear can detect at the frequency concerned. **Sensitivity has its maximum value at a frequency of about 2 kHz.**

4.6 THE STRUCTURE AND FUNCTION OF THE EAR

The ear is divided into three main parts, the outer, middle and inner ears, by three thin membranes – the **tympanic membrane** (or **ear-drum**) between outer and middle, and the **oval window** and the **round window** between middle and inner (Fig. 4.3).

Fig. 4.3
Structure of the ear

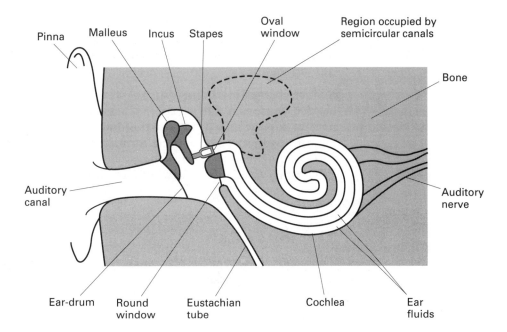

The inner ear contains transducers that convert the mechanical vibrations transmitted to it, via the outer and middle ears, into electrical impulses which then pass to the brain for interpretation.

The Outer Ear

The visible part of the ear (the **pinna**) funnels sound through the **auditory canal** to the ear-drum. The canal is about 2.5 cm long and about 7 mm in diameter. It is closed at one end by the ear-drum, a thin (~ 0.1 mm) membrane with an area of about 65 mm^2, which vibrates with small ($\sim 10^{-11}$ m) amplitude when sound waves enter the ear. The auditory canal acts like a tiny organ pipe with a resonant frequency of about 3 kHz – the frequency at which the ear has its maximum response (see Fig. 4.1).

The Middle Ear (Tympanic Cavity)

This is a small (~ 0.6 cm^3) air-filled chamber containing a chain of three small bones – the **malleus, incus** and **stapes** (or **hammer, anvil** and **stirrup**). They are known collectively as the **ossicles** and they provide a mechanical linkage that transmits the oscillations of the ear-drum to the oval window. (Note. The inertia of the ossicles is such that they cannot vibrate at frequencies in excess of about 20 kHz – hence the ear's upper frequency limit of 20 kHz.) The 'handle' of the hammer is attached to the ear-drum. Its other end is connected, by a synovial joint, to the anvil, and this in turn is connected, by another synovial joint, to the stirrup. Finally, the 'footplate' of the stirrup is attached to the oval window.

The ossicles act as a series of levers with a combined mechanical advantage of 1.3. Furthermore, the area of the ear-drum is 20 times that of the oval window, and therefore the pressure (force / area) at the oval window is $20 \times 1.3 = 26$ times that at the ear-drum. This offsets, to a large extent, the mismatch in the acoustic impedances of the outer and inner ears (see section 8.3).

Tiny muscles attached to the hammer and stirrup allow them, when necessary, to move in such a way as to reduce the amplitudes of oscillation of the ear-drum and the oval window. This action reduces the likelihood of the inner ear being damaged by loud sounds.

Sudden changes in pressure (such as we experience in aeroplanes and lifts) impair hearing, cause pain and, in extreme cases, rupture the ear-drum. To prevent this, a narrow passage called the **Eustachian tube** connects the middle ear to the atmosphere by way of the throat. Its function is to equalize the air pressures on the two sides of the ear-drum. The tube is normally closed but it opens during swallowing, yawning and chewing, and so allows air to enter or leave the middle ear as necessary.

The Inner Ear

This contains the semicircular canals (which are concerned with balance and play no part in the hearing process) and a sensitive, liquid-filled organ called the **cochlea**. The cochlea is connected to the brain by way of the **auditory nerve**.

When the oval window vibrates, a pressure wave propagates through the fluids in the cochlea, stimulating nerve cells and creating the sensation of hearing. Different parts of the cochlea are responsive to different frequencies, and therefore the neurons activated by any particular frequency are specific to that frequency. This allows the brain to distinguish one frequency from another. (The high-frequency receptors are close to the oval window; the low-frequency receptors are deep inside the cochlea.)

4.7 AUDIOMETRY

Hearing threshold levels are determined with a device known as an **audiometer**. This supplies pure-frequency notes, generated electronically, to headphones worn by the patient. The controls allow the operator to alter both the frequency and the intensity level. In a standard audiometric examination, the patient is tested at 8 frequencies from 250 Hz to 8 kHz. The audiometer is calibrated in such a way that a person with normal hearing would register 0 dB at each of the frequencies tested. The volume is gradually increased and the intensity level (in dB) at which the patient can just hear the sound is determined at each frequency in turn. The amount by which this exceeds 0 dB is the hearing loss at that frequency. The results are plotted against frequency to produce an **audiogram**.

Fig. 4.4 compares the audiogram of a person aged 65, whose hearing loss is due solely to ageing, with that of a (younger) person who has been subjected to excessive levels of noise over many years. The curves are typical of the two conditions.

Fig. 4.4
Audiogram showing age-related loss and noise-induced loss

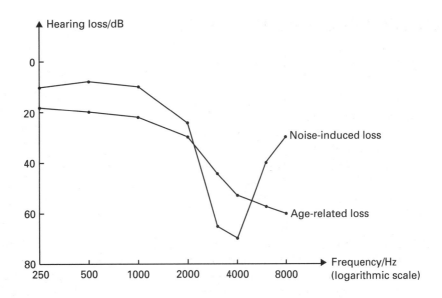

Age-related hearing loss is characterized by a loss in sensitivity across the whole spectrum but which is most marked at the higher frequencies.

Noise-induced hearing loss is characterized by a marked dip in sensitivity around 4 kHz, regardless of the frequencies of the sounds to which the sufferer has been exposed.

CONSOLIDATION

The frequency range of a normal young adult is from 20 Hz to 20 kHz. The upper limit decreases with age. The maximum response (minimum threshold of hearing) is around 3 kHz.

The reference intensity (the minimum intensity the ear can detect) is normally taken to be $1.0 \times 10^{-12}\,\text{W}\,\text{m}^{-2}$.

The ear has a logarithmic response to sound intensity, i.e. increase in loudness is proportional to fractional increase in intensity.

Intensity level takes account of the logarithmic response of the ear. It is measured in decibels (dB). The decibel is used, rather than the bel, because one decibel is (approximately) the smallest change the ear can detect.

Loudness is a subjective quantity. It can be measured in phons and on the dBA scale. Both scales take account of the ear's logarithmic response to intensity and of its unequal response to different frequencies.

Sensitivity is the ability to detect small changes in intensity; it has its maximum value at about 2 kHz.

QUESTIONS ON CHAPTER 4 ▐████████████████████

Assume, where necessary, that the reference intensity $(I_0) = 1.0 \times 10^{-12} \, \text{W m}^{-2}$.

1. (a) Explain what is meant by the *intensity of sound* and indicate how the loudness perceived is related to the intensity received at the ear.

   Calculate the intensity of the loudest sound the ear can withstand given that the intensity level, referred to a threshold for human hearing of $10^{-12} \, \text{W m}^{-2}$, of the same sound is 120 dB.

   (b) The eye has a threshold for perception when 100 photons per second, of wavelength 510 nm, enter the pupil. The effective area of the external entrance of the auditory canal (auditory meatus) is 65 mm².

   If the threshold sensitivity of either organ is defined as the least power required to produce a perceptible signal, calculate the ratio of the threshold sensitivity of the eye to that of the ear.

   Comment on the significance you think this result has for man.

   Planck's constant $= 6.6 \times 10^{-34} \, \text{J s}$
   Speed of light $\quad = 3.0 \times 10^8 \, \text{m s}^{-1}$
   [N, '81]

2. (a) Explain what is meant by the decibel scale for comparing two quantities, and give a definition of a reference level for such a scale for sound intensities.

   (b) A listener wears headphones connected to the output of a stereo amplifier whose output is initially 2.0 mW. The listener slowly increases the output power and subjectively does not discern an increase in sound intensity until the power has risen to 2.5 mW. Successively discernible increases then occur at 3.2 mW and 4.0 mW.

   Use these results
   (i) to show why the decibel scale is a useful one,
   (ii) to calculate the ratio of the amplitudes of the pressure waves of two sounds whose difference in intensity is just discernible. [N, '79]

3. (a) The loudness of a sound as heard by an individual depends on the frequency and intensity of the sound.

   (i) Define *frequency* and *intensity*.
   (ii) Sketch a graph to show how loudness of a sound, as heard by a person with normal hearing, depends on frequency when the intensity of the sound is constant. Point out any special features of the graph.
   (iii) Describe the response of the ear to different intensities of sound with the frequency remaining constant. Hence define *intensity level*.

   (b) A source emits sound energy uniformly in all directions. A sound-level meter records 92 dB when situated 2.5 m from the source. Given that I_0, the threshold intensity of hearing, is $1.0 \times 10^{-12} \, \text{W m}^{-2}$, calculate the total sound power emitted by the source.
   [C, '92]

4. (a) With the aid of a labelled diagram of the middle ear, explain how sound energy is transmitted across the tympanic cavity. Give a reason why the pressure changes due to sound are increased as a result of this transmission and state an approximate value for the increase.

   (b) A meter, which measures relative intensity level of sound referred to $1 \, \text{pW m}^{-2}$, records a value of 97 dB when a pneumatic drill is switched on some distance away. Calculate the intensity of sound at the meter.

   A second drill, identical to the first, is placed close to it. State the intensity of sound at the meter when both drills are working and hence calculate the *increase* in the meter reading.

   Explain why it is convenient to use the decibel scale for such measurements.
   [N, '85]

5. (a) Give a **short** outline of the **physical** processes which give rise to the sensation of hearing when a sound wave is directed towards the human ear.

   (b) The auditory canal along which the sound waves pass can be thought of as being like a pipe closed at one end. In a normal adult the length of this canal is about 0.025 m.

Estimate the fundamental resonant frequency of the canal if the speed of sound in air is $340 \, \text{m s}^{-1}$. Neglect end effects.

(c) (i) Sketch a graph showing how the threshold of hearing of the normal adult human ear varies with frequency when listening to a source of sound of constant intensity over the whole of the audible range.

(ii) Label the axes with appropriate numerical values and units.

(iii) The sound intensity at the threshold of hearing for a frequency of about 3 kHz is $10^{-12} \, \text{W m}^{-2}$. What do you think sets this lower limit?

(d) A foghorn situated at the top of a tall mast may be considered to be a point source of sound. When the horn is in operation the sound intensity measured 0.8 m from the horn is $2.0 \times 10^{-3} \, \text{W m}^{-2}$. An observer sailing towards the horn can just hear it at a distance of 900 m.

(i) What is the sound intensity at the ear of the observer when the sound is first heard?

(ii) Suggest why the intensity in (i) is much larger than the intensity at the threshold of hearing. [N, '93]

6. (a) Define *intensity of sound* and give a formula for relative intensity level measured in decibel (dB) units. Explain why it is convenient to compare intensity levels of sound in decibel units.
The value of the *standard reference intensity* is $10^{-12} \, \text{W m}^{-2}$. Explain what this value represents.

(b) The graph shows the threshold of hearing, for a young person with good hearing, as a function of frequency.

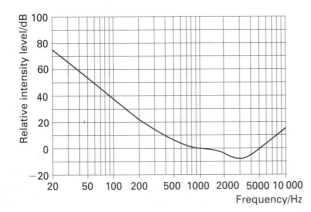

(i) Estimate the frequency at which the subject's ear is most sensitive.

(ii) Estimate the least intensity of sound the subject can detect at 100 Hz.

(iii) The external auditory canal of the subject behaves like a tube of effective length 2.8 cm closed at one end by the eardrum. Show, with the aid of a calculation, that this is consistent with your answer to (i) above.

The speed of sound in air $= 340 \, \text{m s}^{-1}$.
[N, '83]

7. (a) A certain source of sound of variable frequency produces sound of the same intensity at all frequencies.

(i) Sketch a graph, labelled Graph 1, showing the relationship between the response to sound from the source of a sound level meter scaled in dB (vertical axis) and the frequency (horizontal axis) as the latter is varied from 10 Hz to 20 kHz. Indicate numerical values on the frequency axis.

(ii) On the same set of axes sketch a graph, labelled Graph 2, to show the relationship for a meter scaled in dBA as the frequency is varied over the same range.

(iii) Why do most sound level meters use the dBA scale?

(b) A lightning flash is seen some distance away and 4 s later the accompanying thunderclap is heard at a site where a sound level meter produces a maximum reading of 100 dB. Assuming the thunder behaves as if it came from a point source from which the sound intensity falls off with distance according to an inverse square law, calculate

(i) the distance between the lightning stroke and the observation site,

(ii) the peak sound intensity at the meter,

(iii) the peak acoustic power, in watts, produced by the thunderclap.

Neglect absorption of sound during transmission through the air. (Speed of sound in air $= 331 \, \text{m s}^{-1}$.)

(c)

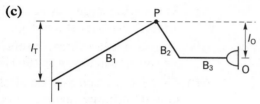

The diagram shows, schematically, a simplified version of the sound transmission system of the middle ear, where T represents the tympanic membrane and O the oval window. Membrane and window are linked by the rigid system of rods, B_1, B_2 and B_3, representing the ossicles. The rods rotate in the plane of the diagram about the pivot P, while the dimensions l_T and l_O are the vertical distances from the pivot to the membrane and window respectively.

(i) Name the three ossicles represented by B_1, B_2 and B_3.

(ii) Find the force, F_O, on the oval window in terms of the force, F_T, acting normally on the tympanic membrane and the given distances.

(iii) Show that, if an incident sound wave produces a pressure, P_T, on the tympanic membrane, the resultant pressure, P_O, on the oval window is given by

$$P_O = P_T \frac{A_T \, l_T}{A_O \, l_O},$$

where A_T = area of tympanic membrane and A_O = area of oval window.   [N, '91]

8. (a) Describe the basic structure of the ear and explain how the ear responds to an incoming sound wave.

(b) The average intensity of sound from a quiet conversation is $5.0 \times 10^{-6}\,\mathrm{W\,m^{-2}}$; the threshold intensity is $1.0 \times 10^{-12}\,\mathrm{W}$ $\mathrm{m^{-2}}$.

Answer the following questions about this statement.

(i) What is meant by the term *intensity* when applied to a sound wave?

(ii) What does the term *threshold intensity* mean?

(iii) How would you measure the *average* intensity, given that you have a sound-level meter?

(iv) What is the intensity level, in dB, of the quiet conversation?   [C, '93]

9. Tables of sound intensities, such as the one which is given here, are often published in textbooks dealing with human hearing.

Sound	Intensity/ Wm^{-2}	Intensity level/ dB
(Rupture of ear-drum)	10^4	160
Jet engine	10	130
(Threshold of pain)	1	120
Loud thunder	10^{-1}	110
Heavy street traffic	10^{-4}	80
Normal conversation	10^{-6}	60
Whisper	10^{-10}	20
(Threshold of hearing)	10^{-12}	0

(a) What is meant by the term *intensity*?

(b) By inspection of the table deduce
  (i) the intensity of the sound from a motor cycle when the intensity level is 100 dB,
  (ii) the intensity level of bird song with an intensity of $10^{-8}\,\mathrm{W\,m^{-2}}$.

(c) The ear of a particular person collects sound from an effective area of 15 cm^2. What power of sound is the ear detecting when at the threshold of hearing?

(d) Explain how the power of the sound entering a person's ear is used to maintain resonant frequency oscillations in the inner ear.

(e) State three factors which make obtaining data for a table such as this, very difficult.

(f) Suggest additional information which needs to be given in order to make use of the values given in the table.   [C, '95]

10. (a) (i) A sound level meter measuring the noise from a tractor in the open air gave a reading of 76 dB. The tractor was then driven into a large hall for an agricultural exhibition and the meter reading rose to 94 dB. Calculate the ratio of the intensity of sound inside the hall to that outside.

  (ii) Explain why the meter used for the measurements should be scaled in dBA.

(b) (i) Sketch a graph showing the threshold of hearing for a young person with good hearing as a function of frequency, marking approximate scales on the axes. Using the same axes, sketch a graph of the threshold of the same person after being exposed for more than a year to noise levels, such as those from the tractor indoors, without wearing ear defenders.

(ii) State how the deterioration of hearing with age differs from that due to exposure to excessive noise.

[N, '89]

11. (a) Give approximate values for the relative intensity levels in dB for:

(i) sound due to normal conversation,

(ii) the threshold of hearing for the ear at its most sensitive in a young person with normal hearing,

(iii) the threshold of pain for such a person.

Give the value of a typical frequency at which the normal ear is most sensitive.

By considering the differences between age-related hearing loss and loss due to exposure to excessive noise, explain how it is possible to estimate the hearing damage due to the latter in a person of given age.

(b) At the site of a new machine in a factory, the relative intensity level measured 70 dB before the machine was brought into operation and 79 dB when it was running. Find the relative intensity level due to the machine alone.

(c) Most sound level meters give readings on a modified dB scale. Give the name of the usual modified scale and explain why it is used instead of relative intensity level in dB.

[N, '87]

12. The middle ear has three bones positioned so as to amplify the pressure changes produced by an incident sound wave.

(a) Why is this amplification necessary?

(b) The diagram illustrates the lever system in the middle ear. Forces F_1 and F_2 are applied over areas A_1 of the ear drum and A_2 of the oval window respectively.

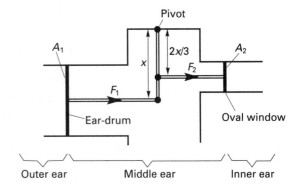

Explain, with reference to the diagram, how two features of the structure of the ear give rise to an amplification of pressure changes.

[C, '95]

5

ELECTRICAL ACTIVITY

5.1 NERVE IMPULSES: ACTION POTENTIAL

Nerve cells (neurons) have long thin extensions known as **nerve fibres (axons)** along which **nerve impulses** are propagated at speeds of up to $100\,\mathrm{m\,s^{-1}}$.

An axon consists of a central core surrounded by a membrane through which small ions (in particular Na^+ and K^+) can pass under certain circumstances. When the axon is at rest (i.e. when it is not conducting an impulse) the fluid inside it (**axoplasm**) contains large, negatively charged, organic ions (most of which are proteins) and a high concentration of K^+ ions; the fluid immediately outside it (body tissue fluid) has a high concentration of Na^+ ions (Fig. 5.1). There is an excess of positive charge on the outside – the membrane is said to be **polarized**, and there is a PD across it of about $70\,\mathrm{mV}$. (It is customary to regard the outside of the membrane as having a potential of zero, in which case the potential inside, the so-called **membrane potential**, is $-70\,\mathrm{mV}$.)

This ionic imbalance is the result of an equilibrium between diffusion and a process called the **sodium–potassium pump**. The 'pump' moves Na^+ ions out of the axon and K^+ ions into it. Diffusion returns some of the K^+ ions to the outside (where their concentration is lower), but is unable to return Na^+ ions to the inside because the membrane is relatively impermeable to (the larger) Na^+ ions.

When the nerve cell is stimulated, the membrane suddenly becomes permeable to Na^+ ions and they are able to move into the axon as a result of both diffusion and the influence of the negative charges inside it. This increases the positive charge inside the axon and so increases the membrane potential, first to zero (known as **depolarization**) and then to $+30\,\mathrm{mV}$ (**reverse polarization**) – see Fig. 5.2. Almost immediately, the membrane becomes impermeable to Na^+ ions and so traps them inside the axon. K^+ ions continue to diffuse out and quickly restore the positive potential outside the membrane (**repolarization**). This sequence of events, which takes only about 2 ms, is known as an **action potential**. A much slower process (lasting about 50 ms) returns the axon to its initial state, with K^+ ions on the inside and Na^+ on the outside, so that it is ready to respond to the next stimulus.

Fig. 5.1
Cross section of a resting
(polarized) nerve fibre

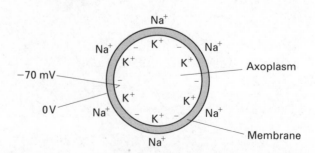

Fig. 5.2
Action potential of a
nerve fibre

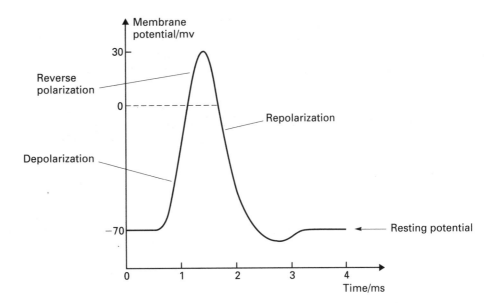

When part of the membrane becomes depolarized it triggers the (still polarized)
part next to it to go through the same sequence of events as itself, (i.e. through the
same action potential). This then triggers the next region, and so on so that the
nerve impulse propagates along the fibre.

5.2 THE HEART

The heart (shown schematically in Fig. 5.3) is a double pump consisting of four
chambers – the right and left **atria** (or **auricles**), and the right and left **ventricles**.
The right-hand chambers take oxygen-depleted blood from the body and pass it to
the lungs; the left-hand chambers take oxygen-rich blood from the lungs and pass
it to the body.

The heart, like any muscle, contracts when subjected to an electrical stimulus. The
stimuli which control the regular beating of the heart are produced by a specialized
group of muscle cells located in the right atrium, and known as the **sinoatrial** (or
SA) **node**. This generates a pulse, about 70 times a minute, which spreads out

Fig. 5.3
The heart (schematic)

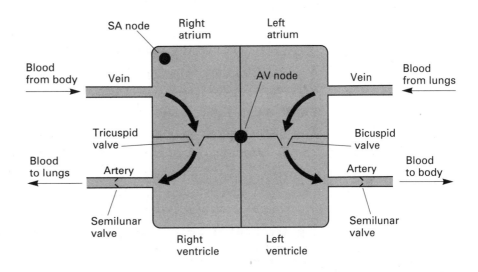

over both atria causing them to contract and force blood into their respective ventricles. The pulse passes to the **atrioventricular** (or **AV**) **node**, where it is delayed for about 0.1 s. It then spreads rapidly across the ventricles causing them to contract and force blood into the arteries and round the body. Finally, the heart relaxes and draws blood in through the veins so that the cycle can start again.

When the ventricles contract, the blood is prevented from returning to the atria by two one-way valves (the **bicuspid** and the **tricuspid**). A second pair of one-way valves in the arteries (the **semilunar valves**) prevents blood being sucked back into the ventricles when they subsequently relax.

The action potential associated with the contraction and relaxation of the heart is shown in Fig. 5.4. Depolarization causes the heart to contract; repolarization causes it to relax. Each complete cycle corresponds to one heartbeat. The membrane potentials, like those of nerve cells, are created by ionic imbalance.

Fig. 5.4
The action potential of
the heart muscle

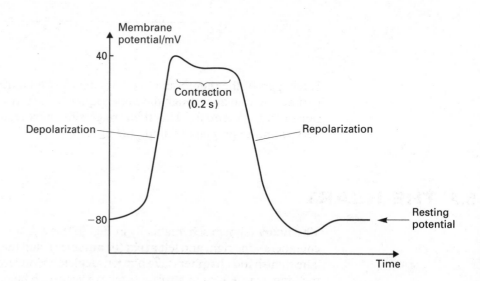

5.3 THE ELECTROCARDIOGRAM (ECG)

The PD which exists between polarized and depolarized heart cells can be detected at the surface of the body. The signals are much attenuated by their passage through body tissue and require amplification before being displayed on an oscilloscope or chart recorder. The display is called an **electrocardiogram** (**ECG**) and it can provide useful information about the condition of the heart.

A typical ECG, covering a single heartbeat of a normal heart, is shown in Fig. 5.5. It has three distinct features.

The P-wave: depolarization and contraction of the atria.

The QRS-wave: depolarization and contraction of the ventricles.

The T-wave: repolarization and relaxation of the ventricles.

Note   There is no wave to show the repolarization of the atria because this occurs at the same time as the large QRS-wave and is masked by it.

The ways in which some common heart conditions affect the trace are listed in Table 5.1.

Fig. 5.5
ECG of a normal heart

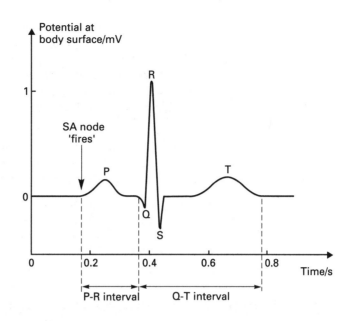

Table 5.1
Some common heart
disorders

Feature	Possible cause
Jagged trace	Ventricular fibrillation – rapid twitching of ventricles with very little actual pumping
Decreased QRS height	Reduced ventricular contraction
Decreased T height	Heart muscle lacks oxygen
Increased T height	Excess potassium in body
Increased Q–T interval	Heart attack
Increased P–R interval	Scarring of atria and/or AV node

5.4 OBTAINING THE ECG

The exact shape of the waveform depends on where the electrodes are placed. There are twelve standard arrangements, none of which involves the right leg because it is too far from the heart. In the so-called bipolar connections, any two of the other three limbs may be used (Table 5.2). Unipolar limb connections use three electrodes, two of which are held at zero potential and are connected to two different limbs. By attaching the third electrode to the remaining limb, it is possible to record the difference in potential between one limb and the average of the other two. The unipolar chest arrangement compares the potential at any one of six sites on the chest close to the heart with the average potential of the three limbs.

Table 5.2
Standard electrode
arrangements

Arrangement	First electrode	Neutral electrode(s)	Name
Bipolar	Right arm	Left arm	Lead I
	Right arm	Left leg	Lead II
	Left arm	Left leg	Lead III
Unipolar limb	Right arm	Left arm, Left leg	aVR
	Left arm	Right arm, Left leg	aVL
	Left leg	Right arm, Left arm	aVF
Unipolar chest	One of 6 chest sites	Both arms, Left leg	V_1 to V_6

In order to reduce contact resistance, hairs and dead cells are removed from the skin at the points where the electrodes are attached, and the skin is smeared with a conductive gel. The electrodes are held in place with adhesive tape – any relative movement of the skin and the electrodes would produce a 'noisy' signal. The electrode material must not react with chemicals produced by the skin. If it did, the electrodes would polarize and block the signal. Silver is commonly used as the electrode material.

The patient should be relaxed, because any anxiety will be reflected in the heartbeat and could lead to a false diagnosis. Movement should be avoided because muscular activity will superimpose unwanted signals on the trace.

A high-gain amplifier is required. The contact resistance is high, despite the presence of the conducting gel, and therefore the amplifier should have a high input impedance ($> 1 \, M\Omega$) to ensure good signal transfer. For safety, the connections to the patient should be electrically isolated from the mains supply. Noise from nearby electrical equipment can be reduced by using a differential amplifier.

5.5 THE ELECTROENCEPHALOGRAM (EEG)

The action potentials of nerve cells in the brain give rise to electrical signals that can be detected at the surface of the skull – **brain waves**. They can be monitored by placing electrodes on the skull. The signals have only small amplitude ($\sim50 \, \mu V$) but after suitable amplification they can be displayed on an oscilloscope or chart recorder. The trace obtained is called an **electroencephalogram** (**EEG**).

The waveform is not repetitive like that produced by the heart, and exhibits many variations in both frequency and amplitude. The brain of a normal person produces four distinct types of wave (Fig. 5.6).

Alpha waves (8–13 Hz) are produced when the mind is relaxed and the eyes are closed.

Beta waves (14–100 Hz) occur during mental activity.

Delta waves (0.5–3.5 Hz) occur during deep sleep.

Theta waves (4–7 Hz) are usually found in children. They also occur in adults suffering emotional stress.

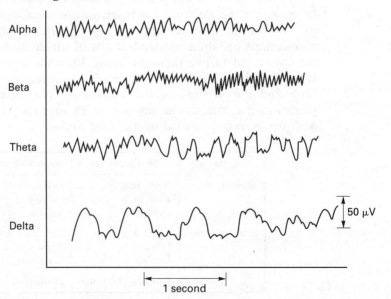

EEGs have been used:

(i)   In the diagnosis of brain disorders such as epilepsy.

(ii)   In research on the nature of sleep.

(iii) To monitor the effects of anaesthesia during surgery.

(iv)   To provide evidence of brain death.

QUESTIONS ON CHAPTER 5

1. Surface electrodes are attached to the right and left arms of a healthy young patient. The potential difference between the two electrodes is recorded continuously using an electrocardiograph.

    (a) Sketch a graph of the observed potential difference as a function of time over one complete heart cycle.

    (b) Label the axes and indicate the typical signal amplitude and time of one complete cycle.

    (c) Why are the electrode surfaces which contact the skin covered with a gel containing a strong electrolyte before being attached to the patient?

    (d) Explain why a differential amplifier is used in processing the signals from the two electrodes.   [N, '93]

2. (a) Sketch a graph of the action potential as a function of time for a typical nerve axon, giving approximate scales on the axes. Air breaks down so that a current flows when the electric field strength is $2.5 \times 10^6\,\mathrm{V\,m^{-1}}$. If a typical axon has a membrane thickness of 10 nm state, giving your reasoning, whether air or the membrane is the better insulator.

    4.3×10^{-8} mol of sodium ions enter the core of an axon per square metre of membrane area during an action potential lasting one millisecond. Calculate the average electric current density associated with this ionic flow and the average electric current if the action potential involves a membrane area of $5.0 \times 10^{-12}\,\mathrm{m^2}$.

    Charge of an electron $= -1.6 \times 10^{-19}\,\mathrm{C}$
    The Avogadro constant $= 6.0 \times 10^{23}\,\mathrm{mol^{-1}}$

    (b) Electrodes are placed on the surface of the body to record the cardiac waveform in a healthy person.

        (i) Sketch a graph of potential difference between the electrodes as a function of time during a single beat of the heart, giving approximate scales on the axes.

        (ii) Mark on the time axis the approximate points when the sino-atrial node is triggered and when ventricular stimulation occurs.

        (iii) What change would you expect to find in the electrocardiogram of a patient suffering from poor ventricular contraction?

        (iv) Why should the person under examination be as quiet and relaxed as possible?   [N, '87]

3. (a) Sketch the waveform produced by an electrocardiograph from electrodes attached to the surface of the chest of a healthy human subject. Label the axes with appropriate values and mark the events in your waveform associated with the following features: ventricular depolarization, ventricular repolarization, atrial repolarization.

    Explain

        (i) the meaning of depolarization and repolarization,

        (ii) the roles played by the atria and the ventricles.

    (b) Two electrodes used to pick up the electrical signal from the chest are each of contact resistance R. They are connected to an amplifier between whose input terminals there is a resistance of $1.0\,\mathrm{M\Omega}$. Calculate the value of R such that the voltage transferred to the amplifier input terminals is 75% of that appearing between the contact points on the skin.

    Name **one** other desirable property of the amplifier apart from large voltage amplification and high input resistance.

    (c) Explain the method used to ensure good electrical contact between the electrode and the skin.

    Why must the subject be relaxed in order to obtain a good electrocardiograph recording?   [N, '90]

6

MEASUREMENTS OF PRESSURE AND TEMPERATURE

6.1 THE SPHYGMOMANOMETER

The sphygmomanometer provides a simple, <u>non-invasive</u> method of measuring blood pressure. It consists of an inflatable cuff connected to a mercury manometer by flexible tubing (Fig. 6.1). The cuff is wrapped around the upper arm at the level of the heart. Air is pumped into the cuff (by repeatedly squeezing the bulb) until the pressure in it is more than sufficient to stop the blood flow in the brachial artery. If a stethoscope is placed on the same artery, below the elbow, no sound will be heard once the blood flow has stopped. The pressure is now gradually reduced by opening the release valve. When the **systolic pressure** (i.e. the maximum arterial pressure) is reached, a series of clicks can be heard, corresponding to the artery being forced open momentarily once each heartbeat. The cuff pressure is further reduced, and eventually a point is reached at which the sounds suddenly become very faint. The external pressure is now equal to the **diastolic pressure** (i.e. the minimum arterial pressure) and the artery is able to remain open for the whole of the cardiac cycle.

Fig. 6.1
The sphygmomanometer

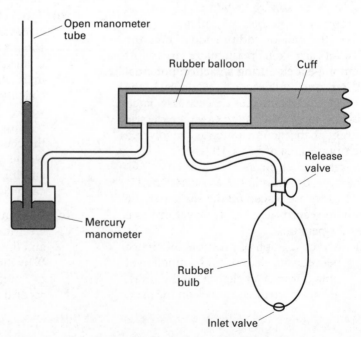

Blood pressure measurements can provide valuable information concerning the condition of the heart and the blood vessels.

Notes (i) For a normal young adult male, systolic pressure = 120 mmHg, diastolic pressure = 80 mmHg. This is expressed as 120/80. The corresponding value for a female is about 110/70.

(ii) The pressures given in (i) are **gauge pressures**. The actual (i.e. the **absolute pressures**) exceed these by atmospheric pressure. Gauge pressure is the pressure read directly from the sphygmomanometer – note that the manometer tube is open to the atmosphere.

(iii) The various sounds that are heard are called **Korotkoff sounds**.

(iv) It is important that the blood supply to the lower arm and hand is restored as quickly as possible.

(v) Although it is **non-invasive** and easy to use, the sphygmomanometer has a number of limitations.

   (a) It does not measure blood pressure directly.
   (b) It is less accurate than invasive methods.
   (c) It measures only the systolic and diastolic pressures; it provides no information on how the pressure changes over the whole of the cardiac cycle.

EXAMPLE 6.1

The average pressure recorded by a sphygmomanometer at the level of a woman's heart is 1.20×10^4 Pa. Assuming that she is in an upright position, calculate the average pressure that would be recorded at a point in her leg 110 cm below the level of her heart. Assume that the pressure difference is due entirely to gravity. (Density of blood = 1.04×10^3 kg m^{-3}, $g = 9.81$ m s^{-2}.)

Solution

Pressure increases with depth. The increase in pressure, Δp, due to a depth, h, in a fluid of density, ρ, is given by

$$\Delta p = h\rho g \tag{6.1}$$

∴   Increase in pressure = $1.10 \times 1.04 \times 10^3 \times 9.81 = 1.12 \times 10^4$ Pa

∴   Pressure in leg = $1.20 \times 10^4 + 1.12 \times 10^4 = 2.32 \times 10^4$ Pa

QUESTIONS 6A

Assume $g = 9.81$ m s^{-2} and density of mercury = 1.36×10^4 kg m^{-3}.

1. Use equation [6.1] to find the atmospheric pressure in pascals (Pa) on a day when the mercury barometer is reading 772 mmHg.

2. Express a pressure of 1.60×10^4 Pa in mmHg.

6.2 PRESSURE MEASUREMENT USING INVASIVE TECHNIQUES

A pressure transducer can be mounted in the end of a **catheter** (a tube) inserted in a blood vessel at the point where the pressure measurement is required. A transducer used in this way is known as an **internal pressure transducer**. The technique is not restricted to the measurement of blood pressure – it can be used to measure fluid pressures in the bladder and the gastrointestinal tract, for example.

External pressure transducers are used in some circumstances. The transducer remains outside the body, connected to the site of the pressure measurement by means of a saline-filled catheter. The pressure inside the chambers of the heart is normally measured in this way.

Various types of pressure transducer are available and some of these will now be discussed. The type chosen for any particular pressure measurement depends on the site of the measurement, how large the pressure is and how rapidly it is fluctuating. The pressure in a vein, for example, is low (0 to 10 mmHg) and changes only slowly; arterial pressure, on the other hand, is much higher (up to 120 mmHg) and fluctuates rapidly.

Resistance (Strain Gauge) Transducers

Four identical strain gauges are bonded to a metal diaphragm in such a way that when the diaphragm distorts under pressure, two are extended and two are compressed (Fig. 6.2). Each strain gauge is a resistance wire, and the resistance either increases or decreases according to whether the gauge has become longer or shorter. (This follows from the well-known relationship $R = \rho L / A$.)

Fig. 6.2
A resistance (strain gauge) transducer

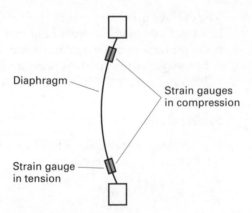

Diaphragm

Strain gauges in compression

Strain gauge in tension

The gauges are normally connected into a bridge circuit as in Fig. 6.3, where R is the unstrained resistance of each gauge and ΔR is the change in resistance. Initially, when all four resistances are the same, the potential at A is equal to that at B and therefore the meter reading is zero. When the diaphragm distorts, the resistances are as shown – the potential at A is now less than that at B, and the meter registers a few millivolts. Alternatively, the meter may be calibrated to read pressure directly.

Using a bridge circuit eliminates the effects of changes in resistance that are due to changes in temperature rather than to changes in length. If the temperature changes, all four resistances are affected to the same extent and no error is introduced.

Fig. 6.3
Bridge circuit after the
diaphragm has distorted

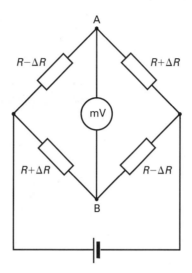

Capacitance Transducers

The capacitance, C, of a parallel-plate capacitor is given by

$$C = \frac{\varepsilon A}{d}$$

where d is the separation of the plates, A is their area of overlap and ε is the permittivity of the material (the dielectric) between the plates. If the dielectric or either of the plates, is attached to a diaphragm (Fig. 6.4), the capacitance will change whenever the diaphragm moves, i.e. whenever the pressure changes.

Fig. 6.4
Capacitance transducers
(a) variable separation (b)
variable area (c) variable
permittivity

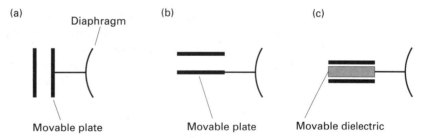

Inductance Transducers

The inductance of an iron-cored coil can be made to respond to changes in pressure by attaching the iron core to a diaphragm in such a way that movement of the diaphragm moves the core in or out of the coil. A commonly used arrangement, known as a **linear variable differential transformer (LVDT)** is shown in Fig. 6.5. The primary induces a voltage in each of the secondary coils. These are connected in opposition to each other and therefore the output is zero when the two secondary voltages are equal, i.e. when the core is centrally placed. If the core moves from this position, one voltage increases and the other decreases, giving an output that varies linearly with the distance the core has moved, i.e. with pressure.

Fig. 6.5
Linear variable
differential transformer

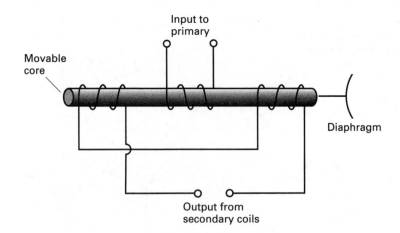

QUESTIONS 6B

1. Refer to Fig. 6.3. By how much does the millivoltmeter reading change when the strain gauge resistances change by 0.20% and the supply voltage (V) is 12 V?

6.3 MERCURY-IN-GLASS CLINICAL THERMOMETER

The clinical thermometer (Fig. 6.6) is a particularly sensitive mercury-in-glass thermometer. It has a very narrow capillary so that large changes in the length of the mercury column occur for only small changes in temperature. The constriction prevents the mercury thread to the right of it returning to the bulb when the thermometer is removed from the patient. The instrument therefore continues to display the correct reading even though it will have cooled before the reading is taken. The mercury is returned to the bulb by giving the thermometer a sharp jerk. The glass in front of the scale is thickened so that it acts as a lens and makes the scale easier to read.

Fig. 6.6
Mercury-in-glass clinical
thermometer

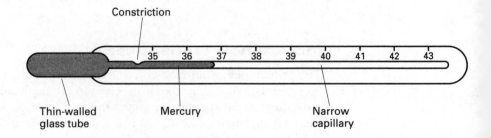

Clinical thermometers are commonly used to estimate body temperature by being placed in the mouth or under the armpit. They cannot be used to monitor rapidly changing temperatures because they have relatively large heat capacities and therefore take a few minutes to reach thermal equilibrium.

6.4 THERMISTORS

Thermistors rely on their change of electrical resistance as a means of measuring temperature. They are semiconducting devices and have negative temperature coefficients of resistance. The resistance decreases approximately exponentially with increasing temperature.

Those used in medicine normally consist of a small bead of semiconducting material, a few tenths of a millimetre in diameter, encapsulated in a thin glass envelope and attached to two connecting wires. They can detect temperature changes of as little as 0.01 °C. They have rapid response times because they have small heat capacities. Their small size allows them to be inserted into blood vessels to monitor blood temperature. More routinely, they are used to measure skin temperature or are inserted in the rectum to measure core temperature. A thermistor is essentially an electrical device and therefore its output can be fed to a chart recorder to provide a continuous recording of temperature.

6.5 THERMOCOUPLES

Whenever two dissimilar metals are in contact an EMF is set up at the point of contact. The magnitude of this EMF depends on the temperature at the junction of the two metals, and therefore the effect (known as the **thermoelectric or Seebeck effect**) can be used in thermometry. The devices which are used in this way are called thermocouples, and at their simplest consist of two wires of different metals joined to each other and to a high-resistance millivoltmeter as shown in Fig. 6.7. The reading on the millivoltmeter increases as the temperature of the junction increases, due to the increased EMF at the junction.

Fig. 6.7
Simple thermocouple

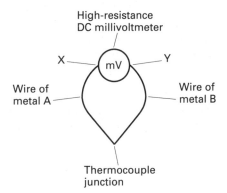

This simple arrangement has a serious disadvantage. Suppose metal A is chromel and metal B is alumel (these two alloys are commonly used in the manufacture of thermocouples), and that the terminal posts of the meter are brass. At X then, there is an EMF due to a chromel/brass thermocouple, and at Y there is a different EMF due to a brass/alumel thermocouple. The meter reading will be the algebraic sum of the three EMFs, and not the EMF of the actual chromel/alumel thermocouple which is required.

This difficulty can be overcome by using a second junction as shown in Fig. 6.8. With this arrangement, the EMFs produced at the meter terminals are equal and opposite, and therefore cancel each other. The extra junction that has been introduced, the so-called 'cold' junction, acts as a reference junction. The hot junction acts as the temperature measuring junction.

Fig. 6.8
Thermocouple with
reference junction

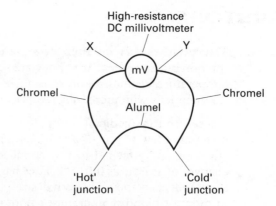

The thermocouple is calibrated on the assumption that the cold junction is in crushed ice and water (i.e. at 0 °C). Since this is not normally convenient, modern devices usually incorporate electronic circuitry to provide automatic cold junction compensation.

Thermocouples have rapid response times. They are used for much the same applications as thermistors and, like thermistors, they can be used with chart recorders to provide a continuous record of temperature.

6.6 THE THERMOPILE

The essential features of a simple thermopile are shown in Fig. 6.9. It can be used to detect the heat radiated by the skin and consists of a number of thermocouples connected in series. One set of thermocouple junctions (A) is exposed to the radiation and is heated by it; the other set (B) is shielded from the radiation. A highly polished metal cone concentrates the radiation on the exposed junctions; these junctions are coated with lamp-black to enhance the efficiency with which the radiation is absorbed.

The meter reading depends on the rate at which heat energy enters the cone and this in its turn depends on the temperature of the skin. Thermopiles are normally calibrated to read skin temperature directly.

Fig. 6.9
Thermopile

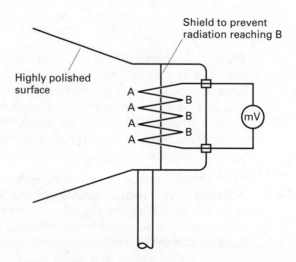

6.7 THERMOGRAPHY

This is the process in which the infrared radiation emitted by the body is used to produce a 'thermal image' or 'temperature map' of the surface of the body. The images are called **thermograms** and are normally displayed on a TV screen. Different temperatures are represented by different colours or, in a black and white display, by different shades of grey.

Thermograph of a man after exertion showing 'hot' hand, forearm and face.

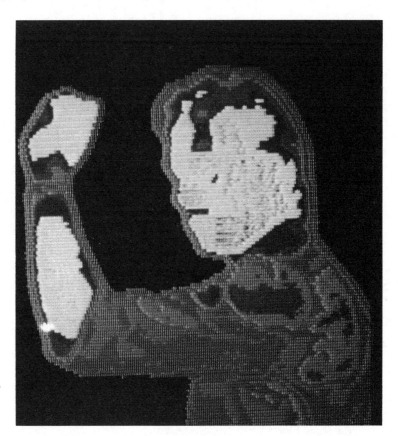

Abnormalities in skin temperature are often evidence of disorders in the underlying tissue. A tumour, for example, is often hotter than the surrounding tissue and shows up as a hot-spot on the thermogram. A blockage in a blood vessel leads to reduced blood flow and a corresponding reduction in temperature.

The Thermographic Camera

Refer to Fig. 6.10. Radiation from a small area of the patient is focused onto an infrared detector – a single crystal of an alloy of indium and antimony whose electrical resistance decreases when infrared radiation falls on it. The circuitry is such that changes in resistance produce changes in the size of the signal output to the amplifier and TV display. The signal-to-noise ratio is increased by cooling the detector in liquid nitrogen to reduce thermal noise.

A two-dimensional image is built up by moving the plane (scanning) mirror so that it sweeps across the patient in a series of horizontal strips, each of which is at a slightly different height from the previous one. The intensity (or colour) of a spot on the screen is determined by the strength of the signal from the detector; its position is determined by the orientation of the scanning mirror.

Fig. 6.10
The thermographic
camera

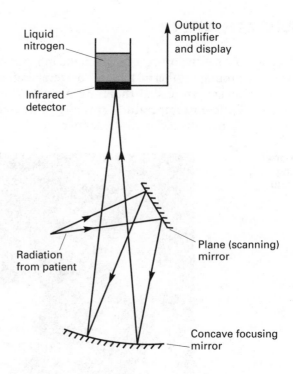

QUESTIONS ON CHAPTER 6

Assume, where necessary, that $g = 10\,\mathrm{m\,s^{-2}}$.

1. **(a) (i)** Draw a labelled diagram of a simple sphygmomanometer showing the three major components of the instrument.

   (ii) Explain how the instrument is used to measure blood pressure. State which pressures are represented by the measurements obtained and give typical values for these pressures in a healthy young adult.

   (b) Arterial blood pressure measurements are made in the head, the foot and at the level of the heart in a healthy young adult. The height of the adult is 1.80 m and the head is 0.50 m above the heart. If the arterial pressure at the heart level is $1.30 \times 10^4\,\mathrm{Pa}$ and any pressure variations in the body are solely due to gravitational effects, estimate the pressure which would be measured

   (i) at the foot when the adult is standing upright,

   (ii) at the head when the adult is standing upright,

   (iii) at the foot when the adult is lying horizontally.

   The adult, again standing upright, is placed in a situation where the acceleration due to gravity is five times greater than the normal value. Estimate how high the blood will rise above the level of the heart. State how this blood level might affect the adult.
   The density of blood may be taken to be $1.05 \times 10^3\,\mathrm{kg\,m^{-3}}$. [N, '94]

2. **(a)** Describe a typical transducer for pressure measurement based on change of electrical resistance, explaining its mode of action.
   A resistance transducer has diameter 1 cm and is to be used to measure blood pressure in an artery. Describe the technique you would expect to be used for such a measurement.
   It is proposed to make a miniature version of the transducer mounted on a catheter tip. If the same materials are used and each linear dimension reduced to one-tenth of its original value, by what factor would the resistance of the sensing element be changed?

   (b) For a person of height 2.00 m, the blood leaving the heart has a mean pressure of 13.30 kPa. If the heart is twice as far from the feet as from the head, calculate the blood pressures in the lower part of the foot and the upper part of the head when the person is erect. Assume gravitational forces alone are responsible for the differences in pressure. (Density of blood $= 1.04 \times 10^3\,\mathrm{kg\,m^{-3}}$.)   [N, '88]

3. **(a) (i)** Describe, with the aid of a diagram, the structure of a pressure transducer based on variable inductance.

    (ii) Describe the changes which occur when pressure is applied to such a transducer and explain how the output is related to the applied pressure.

    (iii) Explain how such an instrument, which is too large to be mounted on a catheter tip, would be used to measure pressure in a blood vessel.

    (b) State the type of instrument you would expect to be most suitable for

    (i) continuous monitoring of the arterial blood pressure of a patient in intensive care,

    (ii) routine measurement of arterial blood pressure during pregnancy.

    (c) A person standing erect on earth has an arterial blood pressure at the heart of 13.6 kPa.

    (i) Calculate the blood pressure you would expect to find at the person's brain, 0.5 m above the heart, assuming gravitational effects alone are responsible for the pressure changes. Assume the density of blood is $10^3\,\mathrm{kg\,m^{-3}}$.

    (ii) Calculate the factor by which the acceleration due to gravity on earth would need to increase if blood were only just to reach the brain.   [N, '91]

7

OPTICAL TECHNIQUES

7.1 OPTICAL FIBRES

An optical fibre consists of a glass core (0.01 to 0.1 mm in diameter) surrounded by a glass cladding of <u>lower</u> refractive index. Most of the light that enters the fibre hits the core/cladding boundary at such an angle that it undergoes **total internal reflection** and so is trapped in the core (Fig. 7.1). It continues to be 'reflected' in this way and eventually emerges from the far end of the fibre.

Fig. 7.1
Total internal reflection in an optical fibre

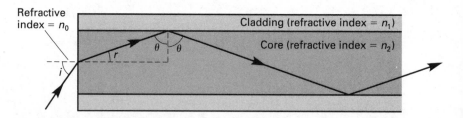

The maximum angle of incidence, i_{max}, for which the light will undergo total internal reflection is given by

$$n_0 \sin i_{max} = (n_2{}^2 - n_1{}^2)^{1/2}$$

where n_1 and n_2 are the (absolute) refractive indices of the core and the cladding respectively, and n_0 is the refractive index of the medium from which the light enters the fibre (usually air).

Proof

Refer to Fig. 7.1. When $i = i_{max}$, $\theta = c$, and therefore $r = (90° - c)$, where c is the **critical angle** for the core/cladding boundary.

$$n_0 \sin i = n_2 \sin r$$

$$\therefore \quad n_0 \sin i_{max} = n_2 \sin (90° - c)$$

$$= n_2 \cos c$$

$$\therefore \quad n_0{}^2 \sin^2 i_{max} = n_2{}^2 \cos^2 c$$

$$= n_2{}^2 (1 - \sin^2 c)$$

$$= n_2{}^2 \left(1 - \frac{n_1{}^2}{n_2{}^2} \right)$$

$$= n_2{}^2 - n_1{}^2$$

i.e. $\quad n_0 \sin i_{max} = (n_2{}^2 - n_1{}^2)^{1/2}$

The quantity $n_0 \sin i_{\max}$ is called the **numerical aperture** of the fibre. Putting $n_1 = 1.42$, $n_2 = 1.49$ and $n_0 = 1.00$ (i.e. air) gives a numerical aperture of about 0.45 and $i_{\max} \approx 27°$, which is typical. In most applications a large numerical aperture is desirable because it allows a wide cone of light to enter the fibre and be transmitted to the other end.

Notes (i)   Bending the fibre decreases θ (see Fig. 7.2) and therefore increases the likelihood that any particular ray will strike the boundary at less than the critical angle.

Fig. 7.2
To illustrate the effect of bending the fibre

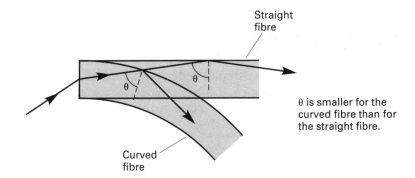

θ is smaller for the curved fibre than for the straight fibre.

(ii) $i_{\max}$ is called the **half-angle** of the fibre because it is half the angle over which the fibre can accept light and transmit it successfully.

(iii)   The reader may be wondering why the cladding glass is necessary. Optical fibres are used in bundles containing tens of thousands of fibres packed closely together. If there were no cladding, light could leak from one fibre to another at points of close contact. The cladding also protects the boundary from mechanical damage.

7.2 TRANSMISSION LOSSES

Light, inevitably, suffers some attenuation in being transmitted along an optical fibre. The main sources of loss are listed below.

(i)   Absorption and scattering by impurities in the core.

(ii)   Losses caused by irregularities at the core/cladding boundary.

(iii)   Partial reflection at the end faces on entry and exit.

(iv)   Losses due to excessive bending of the fibre.

7.3 COHERENT AND NON-COHERENT BUNDLES OF FIBRES

A coherent bundle is one in which the fibres are in the same relative positions at each end of the bundle and which, therefore, can be used to transmit images.

The narrower the individual fibres, and the greater the fraction of the cross-sectional area of the whole bundle that is occupied by core, (rather than cladding)

the greater the **resolution** of the image. The diameter of a single fibre is typically
0.01 mm, and tens of thousands of them are packed together to form a bundle a few
millimetres in diameter.

A non-coherent bundle is one in which the fibres are <u>not</u> in the same relative
positions at each end of the bundle.

They are much cheaper than coherent bundles but can only transmit light – not
images. Since resolution is of no consequence, the fibres can be thicker
(diameter ≈ 0.1 mm), making them more efficient at transmitting light because
there are fewer reflections per unit path length.

7.4 THE FIBRE-OPTIC ENDOSCOPE

This is a flexible tube used for looking inside the body (Fig. 7.3). The shaft of the
instrument is constructed from steel mesh, sheathed in plastic to ease its insertion
into the body. It is about 10 mm in diameter and can be up to 2 m long.

Fig. 7.3
Fibre optic endoscope

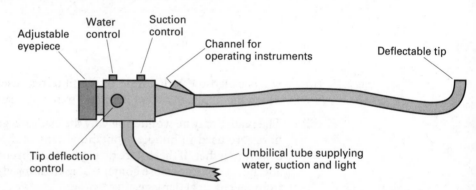

The end inserted into the patient (the **distal end**) can be bent by means of control
wires so that it can be pointed at the area under investigation. The shaft contains
two bundles of fibres – a non-coherent bundle carrying light to provide
illumination, and a coherent bundle which transmits the image back to the
operator. The image is focused onto the end of the bundle by an adjustable
objective lens. An adjustable eyepiece, at the operator end (the **proximal end**),
collects and magnifies the image. The eyepiece normally has provision for a
camera to be attached. The shaft also contains a channel through which water can
be pumped to keep the optics clear, and one by means of which small surgical
instruments can be inserted and manipulated. There may also be a suction
channel to remove any body fluids that are obscuring the field of view.

The **brightness** of the image depends on the intensity of the light source and on
the extent to which light is 'lost' in transmission (see section 7.2). The source is
typically a high-intensity xenon lamp, connected to the proximal end by a second
non-coherent bundle.

7.5 USES OF FIBRE-OPTIC ENDOSCOPY

Endoscopes allow doctors to make visual inspections of inaccessible parts of the
body and, when equipped with the relevant instruments, to carry out minor
surgical procedures. More specifically, they have been used to:

(i)   Provide visual evidence of the presence of ulcers, tumours and internal bleeding.

(ii)   Examine the liver through the wall of the abdomen to look for evidence of cirrhosis – a procedure known as laparoscopy.

(iii) Remove gallstones.

(iv)   Remove foreign bodies from the oesophagus.

(v)   Remove polyps from the colon.

(vi)   Carry out biopsies, i.e. remove suspect tissue to test for malignancy.

(vii) They can also be used in conjunction with lasers – see section 7.6.

7.6 MEDICAL USES OF LASERS

Some of the factors which make the laser a useful medical tool are listed below.

(i)   A laser beam is highly collimated and therefore the energy of the beam can be focused onto a very small ($\sim 10^{-2}$ mm^2) area.

(ii)   Because the beam is highly collimated, it is particularly suitable for transmission along optical fibres. Lasers can therefore be used in conjunction with endoscopes.

(iii) Lasers can cut tissue without causing bleeding – the heat produced by the laser automatically cauterizes the blood vessels as it severs them.

(iv)   There is no physical contact with the tissue being operated on, and consequently less risk of infection.

(v)   A laser beam allows the surgeon an uninterrupted view of the area being operated on. There are situations (operations on the throat, for example) where surgical instruments would block the surgeon's view.

(vi)   Whereas a scalpel tends to damage delicate tissue by dragging it along with the blade, a laser cuts without producing any mechanical distortion.

(vii) The output of a laser is highly monochromatic. Though this is of no advantage in many applications, there are situations in which it allows the surgeon to exploit the fact that different types of tissue absorb different wavelengths to different extents.

Laser eye surgery

Lasers are used to cut, to coagulate and to vaporize unwanted tissue. The type of laser used for any particular application depends on the wavelength of the light it produces, its power output, whether it is pulsed or continuous and whether it can be used with a fibre-optic endoscope. The carbon dioxide laser, for example, is not suitable for fibre transmission because it operates in the far infrared and its output is absorbed by the fibre. Some of the high-power, pulsed lasers are also unsuitable because the sudden temperature increase can shatter the fibre. Typical applications of some commonly used lasers are given in Table 7.1. Carbon dioxide lasers are often used in conjunction with low-power helium–neon lasers. These provide a beam of visible (red) light which is used to align the invisible (infrared) beam of the carbon dioxide laser.

Table 7.1
Examples of laser applications in medicine

Laser	Wavelength	Typical application
Carbon dioxide	Far infrared	Cutting skin and surface tissue Treatment of cervical cancer
Neodymium-YAG	Near infrared	Vaporizing tumours in the gastrointestinal tract Coagulating bleeding stomach ulcers
Argon ion	Blue–green	Repairing detached retinas Removing 'port-wine stain' birthmarks (the blue–green light is readily absorbed by blood)

7.7 SAFETY CONSIDERATIONS WHEN USING LASERS

Because of its ability to focus, the eye is the organ most likely to suffer damage. Anyone entering a laser treatment room should avoid looking directly into the beam and should wear safety goggles designed to protect against the particular wavelengths being employed. Furthermore, the room itself should have as few highly reflecting surfaces as possible. It is not just visible light that presents a danger. The eye can also focus infrared radiation. Not only is this invisible, but it can be strongly reflected by surfaces that are not normally considered to be reflecting.

QUESTIONS ON CHAPTER 7

1. A fibre-optic endoscope contains bundles of optical fibres.
    (a) Explain the difference between coherent and incoherent bundles of fibres.
    (b) Why does the endoscope have to be used in conjunction with a powerful light source?
    (c) Which properties of the fibre bundle affect the ability of the observer to see small details when using the instrument?
    (d) Light falls at an angle of incidence of $15°$ on the plane end of a straight fibre having cladding. The core and cladding have refractive indices of 1.55 and 1.35, respectively. Show, by calculation, whether light will or will not be transmitted down this fibre.

(You may find it helpful to draw a ray diagram showing the passage of a beam of light along the fibre.) [N, '93]

2. **(a)** Describe, with the aid of a diagram, the *optical system* of a flexible fibre-optic endoscope.
    (b)

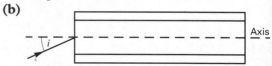

The figure shows an optical fibre consisting of a straight cylindrical core of polymeric material of refractive index 1.49 clad in another polymer of refractive index 1.42.

The ends of the cylinder are normal to the axis. A ray of light is shown incident on the end face of the core at an angle of incidence i, in a plane which contains the axis of the cylinder.

(i) Sketch a ray diagram to show how such light can be transmitted to the other end of the fibre.

(ii) Show that the ray of light will strike the core–cladding interface at an angle of incidence θ, where

$$\sin i = n_{\text{core}} \cos \theta,$$

n_{core} being the refractive index of the core relative to air.

(iii) If θ_c is the smallest angle of incidence on the core-cladding boundary at which total internal reflection can occur, show that

$$\cos^2 \theta_c = 1 - \left(\frac{n_{\text{clad}}}{n_{\text{core}}}\right)^2,$$

where n_{clad} is the refractive index of the cladding relative to air.

(iv) Hence show that the maximum angle of incidence i_m on the end face of the fibre, which allows transmission down the fibre, is given by

$$\sin^2 i_m = (n_{\text{core}})^2 - (n_{\text{clad}})^2.$$

(v) Calculate the maximum angle of incidence for transmission to take place.

(vi) State what purpose the cladding serves in a fibre optic bundle.
[N, '89]

3. (a) (i) Explain what is meant, in fibre optics, by an *incoherent bundle* and by a *coherent bundle*.

(ii) Explain what is meant by an *endoscope*.

(iii) Describe, with the aid of a diagram, a fibre-optic flexible *endoscope* and explain the purposes of its chief parts.

(b) (i) The diagram below, which is not to scale, shows a cylindrical optical fibre, diameter 0.20 mm, of refractive index 1.5, surrounded by air. A ray of light travelling along the axis of the straight portion reaches the region where the axis of the fibre is curved into an arc of radius R. The ray will be reflected at the surface and remain in the fibre provided the angle of incidence exceeds a critical angle i_c where $\sin i_c$ is the reciprocal of the refractive index. Calculate the smallest value of R which allows the ray to remain within the fibre.

(ii) Describe how a fibre with cladding differs from that described in **(b) (i)** and give **one** reason why, in making an optical fibre bundle, fibres with cladding are preferred.

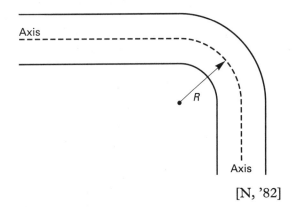

[N, '82]

4. (a) Give **two** examples of the use in medicine of lasers, indicating the properties which make them suitable.

(b) What precautions should be taken to protect the eyes when lasers are used? Give reasons for your answers. [N, '81]

8

ULTRASONIC IMAGING

8.1 INTRODUCTION

Ultrasonic waves (**ultrasound**) are sound waves with frequencies above those that can be heard by the human ear, i.e. above about 20 kHz. The frequencies used for medical applications are usually in the range 1 to 15 MHz.

8.2 THE PIEZOELECTRIC TRANSDUCER

A piezoelectric crystal is a crystal that changes size when a PD is applied to it. If the applied PD is <u>alternating</u>, the crystal vibrates at the frequency of the PD. If the frequency is high enough, the vibrations generate ultrasound. The effect is most marked when the frequency of the applied PD is equal to a natural frequency of vibration of the crystal, i.e. at **resonance**.

The opposite effect also occurs. Thus, if a beam of ultrasound is incident on a piezoelectric crystal, the pressure variations cause the crystal to vibrate and <u>generate</u> an alternating PD.

It follows that a piezoelectric crystal can be used both to generate and to detect ultrasound – this is the basis of the piezoelectric transducer (Fig. 8.1), the same transducer often performing both functions using only a single crystal. A thin disc of piezoelectric material with thin-film, silver electrodes on opposite faces is mounted at the end of a probe. One electrode is connected to the case of the probe and is earthed; the other is connected by a coaxial cable to a power supply when it is acting as a transmitter, and to an amplifier and cathode-ray display when it is acting as a detector. In order to obtain maximum benefit from the resonance effect, the thickness of the crystal is equal to half the wavelength that corresponds to the fundamental frequency of the crystal – each face is an antinode.

Fig. 8.1
The piezoelectric probe

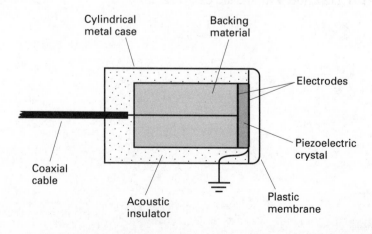

Medical diagnosis using ultrasound is based on the echo-sounding principle. Distinct echoes are obtained by using short (~1 μs) pulses of ultrasound. To facilitate this, a block of epoxy resin, which damps out the vibrations within a few microseconds, is bonded to the back of the crystal.

Note Examples of piezoelectric materials include **quartz** (one of the first to be discovered) and lead zirconate titanate (**PZT**) – a ceramic commonly used in ultrasound transducers.

8.3 REFLECTION OF ULTRASOUND

The specific acoustic impedance (Z) of a material is defined by

$$Z = \rho c$$

where

$$\rho = \text{the density of the material (kg m}^{-3})$$

$$c = \text{the speed of sound in the material (m s}^{-1})$$

Sound is reflected at the boundaries between materials that have different specific acoustic impedances. The ratio of the reflected intensity (I_r) to the incident intensity (I_i) is called the **intensity reflection coefficient** (α). At the boundary between media whose acoustic impedances are Z_1 and Z_2

$$\alpha = \frac{I_r}{I_i} = \left(\frac{Z_2 - Z_1}{Z_2 + Z_1}\right)^2$$

Note, in particular, that:

(i)   When $Z_1 = Z_2$ there is no reflection.

(ii)   When Z_1 and Z_2 are very different most of the incident energy is reflected.

Table 8.1 lists the acoustic impedances of materials found in the body. It is a simple matter to show that a muscle/fat interface reflects only 1% ($\alpha = 0.01$) of the incident ultrasound. This, however, is easily detectable and it means that the rest of the beam can penetrate deeper into the body and produce more images. An air/soft tissue interface, on the other hand, reflects 99.9% of the incident energy. This makes it necessary to use a **coupling agent** (e.g. a gel with an impedance close to that of body tissue) to exclude air between the probe and the skin so that the beam can enter the body.

Table 8.1
Specific acoustic
impedances of some
biological materials

Material	Z/kg m^{-2} s^{-1}
Air	4.29×10^2
Fat	1.38×10^6
Brain	1.58×10^6
Soft tissue	1.63×10^6
Muscle	1.70×10^6
Skull bone	7.78×10^6

8.4 THE A-SCAN

Refer to Fig. 8.2. The transducer is triggered to emit a short (~1 μs) pulse o
ultrasound at the same time as the spot on the oscilloscope starts to move acros
the screen. The pulse enters the body and a series of echoes is reflected back tc
the transducer from various interfaces. Each echo generates a small voltage ot
reaching the transducer. The signals are amplified and fed to the Y-plates of the
oscilloscope, producing a series of voltage spikes on the screen (Fig. 8.3). The
further a spike is from the start of the trace, the deeper inside the body is the
interface that produced it. If the speed of sound in the different types of body
tissue and bone is known, the depth of each interface can be determined. (Note
The speed of sound in soft tissue is about $1500 \, \mathrm{m \, s^{-1}}$; the speed in bone is mucl
greater – about $4000 \, \mathrm{m \, s^{-1}}$.)

Fig. 8.2
The A-scan system

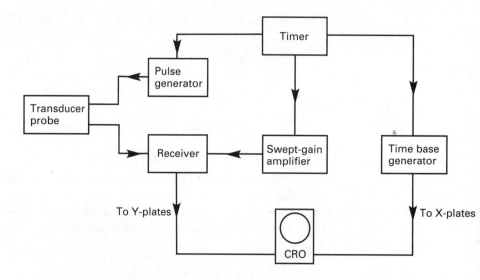

Fig. 8.3
A-scanning

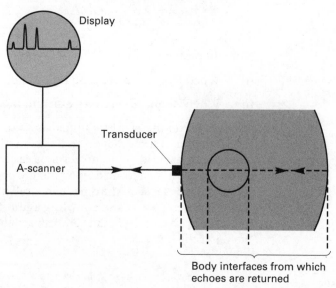

In order to produce an (apparently) continuous trace on the screen a whole series
of pulses is transmitted. The rate at which they are transmitted is known as the
pulse repetition frequency (PRF) and is typically a few kHz, i.e. low enough to
allow time for all the echoes from one pulse to be returned before the next pulse is
sent out.

Echoes from deep inside the body suffer more attenuation than those from near the surface. A **swept-gain amplifier** compensates for this by amplifying the deeper signals to a greater extent than those from nearer the surface.

The A-scan technique is commonly used to measure the size of the foetal skull. It is also used to provide detailed measurements of the eye.

8.5 THE B-SCAN

The B-scan technique produces a two-dimensional picture of a cross-section of the body. Two types of B-scanner are currently in common use; we shall describe just one of these – the **linear array scanner** (Fig. 8.4). This consists of a line of about 100 transducers that are triggered one after another in rapid succession. Each transducer detects the echoes from the various interfaces directly in line with it, and each echo gives rise to a bright spot on the display. The brightness of the spot is determined by the amplitude of the echo. Its position on the screen corresponds to the point in the body that produced the echo. This is achieved by relating the y-component of the position of the spot to the time interval between transmission and reception of the pulse, and the x-component to the position of the particular transducer that produces it. The images are produced in such rapid succession that the picture is created almost instantaneously, and a whole series of complete pictures can be produced one after another (about 25 times per second) to create a 'real-time', moving image.

Fig. 8.4
Linear array scanner

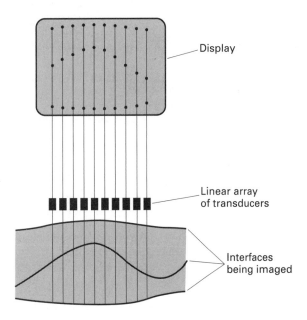

Display

Linear array
of transducers

Interfaces
being imaged

Two factors limit the quality of the image.

(i)   There are gaps in the image corresponding to the gaps between the various transducers.

(ii)   In order for a beam to return to the transducer that transmitted it, it must hit the reflecting interface at 90°. This is clearly impossible when complex surfaces are involved and produces further degradation.

The earliest types of B-scanner had only a single transducer and this was moved across the surface of the body by the operator. The transducer was mounted on a position-sensing arm that allowed the positions of the spots on the display to be correlated with the corresponding points in the body. It took several seconds to build up the image and was therefore of little use for studying parts of the body that were moving.

B-scans are routinely used to monitor foetal development, and to look for cysts, tumours and other abnormalities in various organs of the body. They also allow surgeons to 'see' inside the body, enabling them to carry out operations through small incisions in the skin.

Ultrasound image of human foetus after 21 weeks in the womb

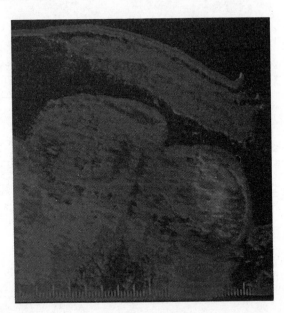

8.6 ATTENUATION

The intensity of a sound wave of a single frequency decreases exponentially with distance as it passes through any given medium. This **attenuation** is due to processes such as absorption, scattering and diffraction. The intensity decreases from I_0 to I in traversing a distance x, where

$$I = I_0 \, e^{-\mu x}$$

The value of μ (the **attenuation coefficient**) depends on both the material concerned and the frequency of the sound. It actually increases with frequency and therefore the **higher the frequency, the greater the attenuation**.

8.7 RESOLUTION

Axial resolution is the resolution in the direction of the beam. **The prime factor in determining axial resolution is the duration of the ultrasound pulse.** A $1\,\mu s$ pulse, for example, travelling through soft tissue at $1540\,\mathrm{m\,s^{-1}}$ occupies a distance of $1.54\,\mathrm{mm}$ and therefore cannot resolve interfaces that are separated by less than this distance. In practice the pulse needs to contain a minimum of about three complete cycles of ultrasound. It follows that higher frequencies allow shorter pulses and therefore increased axial resolution. Unfortunately, increasing the frequency also increases the attenuation (section 8.6) and therefore the optimum resolution is obtained by using the highest frequency that will provide the necessary penetration. Thus, whereas frequencies of 10 MHz or more can be employed for an ultrasound scan of the eye, for example, considerably lower frequencies (5 MHz or less) must be used for investigations deeper inside the body.

Note   In general, because of diffraction, it is not possible to resolve two objects whose separation is less than the wavelength of the waves being used. Ultrasound with a frequency of 1 MHz has a wavelength of 1.54 mm in soft tissue. It follows that the resolution attainable with a 1 μs pulse is not limited by diffraction effects providing the frequency is greater than 1 MHz.

Lateral resolution is the resolution perpendicular to the beam and though it is of little consequence in A-scanning, it is of major importance in B-scanning techniques.

When an ultrasound beam is scanned across a <u>point</u> object, an echo is reflected back to the probe for the whole of the time that the object is in the beam. The image of the object is not a point, but a line whose length is equal to the width of the beam. It follows that two (point) objects will not be resolved if their separation is less than the width of the beam. Thus, **the prime factor in determining lateral resolution is beam width**.

The width of the beam is not constant along its length; at any particular tissue depth it depends on the diameter of the probe and the frequency of the ultrasound. Decreasing the diameter of the probe decreases the width of the beam <u>in the region close to the probe</u>. Unfortunately, the smaller the diameter, the shorter this region becomes. It turns out that this can be offset by increasing the frequency, but this also increases the attenuation. Thus, the optimum lateral resolution is achieved by choosing as small a diameter and as high a frequency compatible with the degree of attenuation that can be tolerated.

Note   The lateral resolution at any particular frequency is typically five times worse than the corresponding axial resolution.

QUESTIONS 8A

1. Calculate the axial resolution at **(a)** 1.0 MHz, **(b)** 5.0 MHz and **(c)** 10 MHz on the assumption that each ultrasound pulse must contain three complete cycles and has an average speed of 1500 m s^{-1}.

8.8 DOPPLER TECHNIQUES

When sound waves are reflected from a <u>moving</u> object, their frequency changes by an amount Δf given by

$$\Delta f = \frac{2fv}{c} \quad \text{(Providing } v \ll c\text{)}$$ [8.1]

where

f = the frequency of the incident waves

v = the velocity of the object relative to the source

c = the velocity of sound in the medium concerned

This is an example of the **Doppler effect** and it provides a non-invasive means of making measurements of **foetal heart movement** and **blood flow** using ultrasound.

The ultrasound beams used in this way are normally continuous wave, rather than pulsed, and therefore the probe has a separate transmitter and receiver (Fig. 8.5). The incident and reflected waves are mixed electronically to produce **beats**. The beat frequency is equal to Δf. This is measured electronically and fed to a display that can be calibrated in terms of v.

Fig. 8.5
Doppler scan system

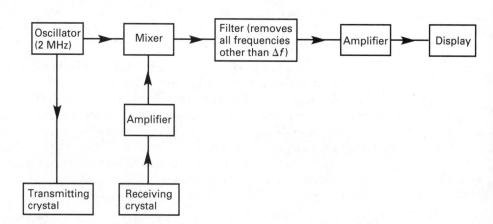

Notes

(i) Δf is positive when the object is moving towards the source of the ultrasound and negative when it is moving away.

(ii) Equation [8.1] assumes that the object whose speed is being measured is moving directly towards the source of the ultrasound. This is not normally the case with blood flow measurements and equation [8.1] needs to be replaced by

$$\Delta f = \frac{2f \, v \cos \theta}{c}$$

where $v \cos \theta$ is the component of the blood velocity in the direction of the reflected beam (see Fig. 8.6).

(iii) Blood flow measurements exhibit a range of Doppler shifts because blood is viscous and therefore flows more quickly in the centre of a blood vessel than near the walls.

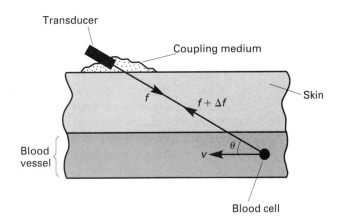

Fig. 8.6
Blood flow measurement

8.9 ADVANTAGES OF ULTRASOUND IN DIAGNOSIS

(i)   It is non-invasive and therefore there is no risk of infection and the patient is not subjected to stress.

(ii) It is safe. The low-intensity ($<0.1\,\text{W m}^{-2}$) beams used for diagnosis are not known to produce any undesirable side-effects.

(iii) It is more effective than (conventional) X-ray techniques in producing images of soft tissue and in revealing the presence of some forms of cancer.

(iv) The equipment is relatively cheap.

8.10 LIMITATIONS OF ULTRASOUND

(i)   A gas/soft tissue interface reflects 99.9% of the incident ultrasound energy. This makes it impossible to obtain images of structures that lie on the far side of the lungs, for example.

(ii) Bone is not easily penetrated by ultrasound and it is therefore difficult to obtain good ultrasound images of the brain. (This is not a problem with the immature skull bone of a developing foetus.)

8.11 ULTRASOUND THERAPY

Ultrasound (at intensity levels of a few W cm^{-2}) has been used to promote the healing of wounds and to provide relief from the pain associated with conditions such as arthritis and fibrositis. The process is not fully understood but it is thought to be connected with the heat produced by the absorption of the ultrasound energy and with the increased blood flow the pressure wave creates – a sort of high-frequency massage.

High-intensity ($\sim 1000\,\text{W cm}^{-2}$) ultrasound beams have been used to break up kidney stones into pieces small enough to be passed out of the body, and so avoid the need for surgery.

8.12 SAFETY ASPECTS

No known hazards are associated with the low-intensity beams used for diagnosis. The higher-energy beams employed for therapy must be used with care in order to avoid excessive increases in tissue temperature. Bone is particularly at risk because it is a strong absorber of ultrasound. The pressure changes associated with these higher-energy beams can rupture cell walls and cause the cells to die. This is not normally a serious problem in adults because only a few cells are likely to be involved, but it is something that should not be risked with a developing foetus, particularly in the early stages of pregnancy.

CONSOLIDATION

Ultrasonic waves have frequencies above 20 kHz. Frequencies from 1 to 15 MHz are used for medical purposes.

Piezoelectric crystals are used both to generate and to detect ultrasound.

Sound is reflected at the boundaries between materials that have different **specific acoustic impedances**. The greater the difference, the greater the reflection.

Coupling agents are used to exclude air between the probe and the skin because an air/skin interface reflects 99.9% of the incident energy.

The A-scan uses the echo-sounding principle to determine the depths within the body of different interfaces.

The B-scan produces a two-dimensional image of a cross-section of the body.

The attenuation of a sound wave increases with increasing frequency.

Axial resolution is improved by using shorter pulses.

Lateral resolution is improved by using narrower beams.

Axial and lateral resolution are both maximized by using the highest frequency that is compatible with the degree of penetration required.

QUESTIONS ON CHAPTER 8

1. **(a)** What range of ultrasonic frequencies is used in medical diagnosis?
   Explain why ultrasonic wavelengths of 0.3 mm or less must be used if we wish to 'see' detail in soft tissue down to 0.3 mm. Calculate the frequency of this ultrasound and state whether it is a maximum or a minimum for resolving these small details.
   (Speed of ultrasound in soft tissue $= 1560 \, \text{m s}^{-1}$.)

   (b) Why is it necessary to limit the power of the ultrasonic source used in diagnosis?

   [L, '92]

2. **(a)** Describe and give a simple explanation of the *piezoelectric effect*.

   (b) A piezoelectric transducer is to be used in a medical application to transmit a repetitive train of short pulses of ultrasound into a body. Give a diagram of, and discuss the features of, such a transducer.

   (c) Estimate the maximum repetition rate which could be used in a transducer which is used as both transmitter and receiver if a body of maximum thickness 0.35 m is to be studied. The speed of ultrasound in soft tissue is about

$1500 \, \text{m s}^{-1}$. Will the rate you have estimated still allow the full thickness of the body to be studied if some bone is encountered?   [O & C, '92]

3. When sound is transmitted from one medium (medium 1) to another (medium 2), the ratio, R, of the reflected intensity to the incident intensity is given by

$$R = \left(\frac{Z_2 - Z_1}{Z_2 + Z_1}\right)^2$$

where Z_2 is the acoustic impedance of medium 2 and Z_1 is that of medium 1.

In the use of ultrasound for medical diagnosis, a coupling medium such as a water-based cellulose jelly is used between the ultrasonic transducer and the patient's skin. Explain why this is so.

(Acoustic impedance of air = 0.430×10^3 $\text{kg m}^{-2}\text{s}^{-1}$,

acoustic impedance of water-based jelly = $1500 \times 10^3 \, \text{kg m}^{-2}\text{s}^{-1}$,

acoustic impedance of tissue = 1630×10^3 $\text{kg m}^{-2}\text{s}^{-1}$.)

Explain the basic principles behind the ultrasound method of obtaining diagnostic information about the depths of structures within a patient's body. Illustrate your answer by reference to a block diagram of a simple (A-scan) scanning system.

What factors limit the quality of the information obtained?   [L, '88]

4. (a) Explain the principles of one method of generation of ultrasonic waves for use in medical diagnosis.

(b) Compare the use of ultrasound with the use of X-rays in medical diagnosis.
                                                [C, '92]

5. A single transducer emitting short pulses of ultrasound is placed in contact with the body so that almost all the energy passes into the body. Some of the energy will be reflected back from inside the body to be detected by the same transducer.

(a) Describe **two** factors which affect the magnitude of the reflected signal.

(b) The transducer may be mounted on a mechanical arm which allows it to be moved over the body surface. Explain how a two-dimensional image, of varying brightness, of a part of the body can be produced using a storage oscilloscope as a signal recorder.

(c) State the advantage of using an array of transducers instead of a single transducer when producing a two-dimensional image.   [N, '94]

6. (a) With the help of a sketch of a typical transducer describe the generation of ultrasonic waves for medical diagnostic purposes.

Explain

(i) why it is essential to use short pulses of ultrasound,

(ii) how it is ensured that the sound energy enters the body,

(iii) the means employed for detection of the received signals.

(b) For pulses of a given power emitted by the transducer mention **two** factors which affect the intensity of the received ultrasound signals.

(c) If the transducer were to be part of a B-scanner, what other fact would you need to know about the received ultrasonic signals? Apart from such data about the received signals, state the additional information needed by the B-scanner if an image is to be produced.

                                                [N, '86]

7. Two diagnostic techniques using ultrasound are called the A-scan system (or A-scope) and the B-scan system (or B-scan imaging). The diagram below shows simplified oscilloscope displays from these two systems.

(a)

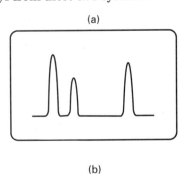

(b)

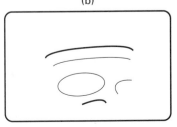

Copy the displays and use them to help you to describe how the main features of
(a) the A-scan display, and
(b) the B-scan image
are produced. (Do not attempt to draw a block diagram for either system.) Your answers should refer to the ultrasonic transducer probe and to the way in which its output is connected to the oscilloscope.

State one diagnostic situation for which each type of scan is useful.   [L, '94]

8. Explain why it is important that when mechanical waves are used in medical diagnosis, the waves used are longitudinal waves. State the range of frequencies usually employed.

What is meant by the Doppler effect in sound? Illustrate your answer with a diagram showing wavefronts.

State two specific applications for the Doppler effect in medical diagnosis.

A beam of ultrasound of frequency 2.0 MHz is being used to measure the speed of a moving surface in a medium where the sound travels at a speed of 1500 m s^{-1}. If the smallest Doppler shift which can usefully be detected is 40 Hz, calculate the minimum measurable speed, v_{min}, of the surface. Explain why other surfaces which are moving faster than v_{min} might not be detected.   [L, '91]

9. A transducer transmits ultrasonic pulses towards a moving surface inside a patient's body. When the transducer detects the signals reflected from this surface they are found to have undergone a 'double' Doppler shift in frequency. Explain why this is so. (Detailed derivations of expressions for the Doppler shift are not required.)

Ultrasound of frequency 5.0 MHz reflected from red blood cells flowing in an artery was found to be Doppler shifted in frequency by 1.50 kHz when the blood flow was at 30° to the direction of propagation of the sound waves. If the speed of the ultrasound is 1.50 km s^{-1}, calculate the speed of blood flow in the artery.

(The expression $\Delta f \approx \dfrac{2fv\cos\theta}{c}$, where the symbols have their usual meaning, may be assumed without proof.)

Why, in practice, is a range of Doppler shifts detected?

Give one advantage of this method of measuring the speed of blood flow.   [L, '88]

10. Draw a block diagram to show the essential components of a continuous wave Doppler system as used for measuring the velocity of moving surfaces within the body.

A 2.2 MHz ultra-sound beam travels at a speed of 1.5 km s^{-1} through soft tissue, and is reflected normally from a moving surface. A Doppler shift of 400 Hz is detected. Calculate the speed of the moving surface.

Explain why ultrasound of high frequency (~10 MHz) is used for scanning an eye, but a lower frequency (~3 MHz) is more suitable for abdominal scans.   [L, '93]

9

MAGNETIC RESONANCE IMAGING

9.1 INTRODUCTION

Magnetic resonance imaging (**MRI**) is a diagnostic procedure that makes use of the phenomenon of **nuclear magnetic resonance** (**NMR**) to produce detailed images of slices through the body. Although NMR can be used to detect the presence of many different species of nucleus, the only nucleus of any interest in MRI is the hydrogen nucleus – because of its high abundance in body tissue.

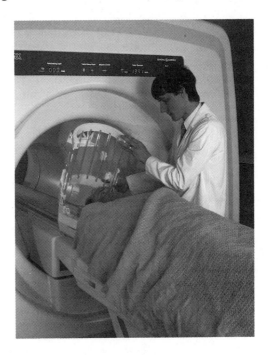

A patient being prepared for a magnetic resonance imaging (MRI) brain scan

9.2 NUCLEAR MAGNETIC RESONANCE (NMR)

Hydrogen nuclei (protons), like all nuclei which do not have an even number of protons <u>and</u> an even number of neutrons, have a magnetic moment and are therefore able to interact with a magnetic field. If a sufficiently strong field is applied, they orient their magnetic moments so that they are either parallel or antiparallel to the field. Those in the parallel state have less energy than those in the antiparallel state. If they are supplied with a **radio-frequency** (**RF**) **pulse** of electromagnetic radiation of a particular frequency, some of them will be induced to 'flip' into the antiparallel orientation. **The frequency depends on the flux density of the applied magnetic field**. (Protons in a field of 2.00 tesla (which is typical) require a frequency of 85.2 MHz.) They return to the parallel orientation once the pulse has passed, emitting radio-frequency signals in the process. These are the NMR signals, and it is these that are used to create the image. (Note. In practice the RF pulse 'flips' the protons in both directions, i.e. parallel to antiparallel and antiparallel to parallel. Strictly, it is the net change that is from the parallel state to the antiparallel state.)

9.3 THE MR SCANNER

An MR scanner requires a magnet that produces a strong field (up to 2 tesla) which is uniform (to better than 1 in 10^6) over a region large enough to accommodate the body section being imaged. **Superconducting magnets*** are normally used. Although these consume no power, the running costs are high because they have to be cooled with liquid helium, which is expensive and which has to be replaced periodically because it gradually boils off.[†] Conventional electromagnets, on the other hand, consume about 40 kW of electricity and are often even more expensive to run.

An **RF transceiver** is used both to transmit the RF pulses that 'flip' the protons and to receive the NMR signals they produce when they return to their normal orientation. The NMR signals are decoded by a computer to obtain the values of two parameters associated with the speed at which the protons return to their normal states. These are called the **T1 and T2 relaxation times**. Different types of tissue are associated with different relaxation times. The image is produced by ascribing different colours (or different degrees of greyness) to different relaxation times.

In order to create the image, it is necessary to locate the point of origin of each of the NMR signals. This is done by using **gradient coils** to superimpose small, non-uniform fields on the main field – effectively, in such a way as to produce a unique field strength at every point in the section being imaged. Since the frequency required to 'flip' the protons depends on the strength of the field, different frequencies correspond to different points in the patient.

9.4 ADVANTAGES AND DISADVANTAGES OF MRI

Advantages

(i)   It does not involve ionizing radiation and therefore does not involve the risks associated with X-ray and γ-ray imaging. As far as is known, there are no harmful effects of any kind. (This last statement assumes that sensible precautions are taken – MRI must not be used on patients who have artificial pacemakers for example.)

---

* A superconductor is a material whose electrical resistivity is zero when it is below a temperature known as its critical temperature.
† Some materials are superconducting at liquid nitrogen temperatures. Those discovered so far are not suitable for magnets – if they were, there would be considerable cost savings because these temperatures are more easily attainable than liquid helium temperatures.

(ii)   There are many situations (e.g. producing images of the brain) in which its ability to distinguish different types of soft tissue is superior to that of CT scanning (section 10.14).

(iii)   It is non-invasive.

Computer-enhanced magnetic resonance imaging (MRI) scan of a human head

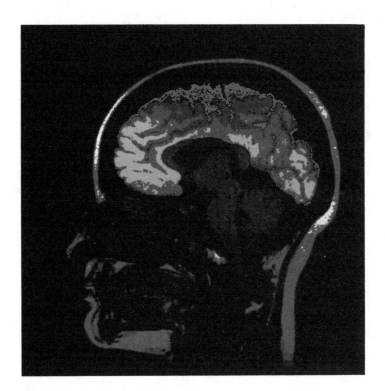

Disadvantages

The major disadvantage is that of cost – MRI is the most expensive of the various imaging processes. High costs are involved both in purchasing the equipment and in running it. Furthermore, the signals that form the image are extremely weak and the cost of screening the scanner from extraneous signals adds significantly to the initial expenditure.

There are a number of other considerations. Careful attention must be given to the position of the scanner because the powerful magnetic field can have an adverse effect on nearby electronic equipment. Personnel need to take care that small metal objects, such as scissors and screwdrivers, are not 'sucked' into the magnet. If they were, they could cause serious damage to the magnet itself or to anyone who happened to be in the way.

10
X-RAYS

10.1 INTRODUCTION

X-rays are high-photon energy (i.e. short-wavelength) electromagnetic radiation. They are used in medicine both for diagnosis (**radiography**) and for therapy (**radiotherapy**). It is not possible either to reflect or to refract X-rays and therefore **X-rays cannot be focused**.

10.2 PRODUCTION OF X-RAYS

In a conventional X-ray tube electrons are emitted thermionically from a heated filament and are accelerated through a high PD towards a metal target (the anode). On colliding with the target, the electrons decelerate rapidly and X-rays are produced. About 99% of the energy of the electrons goes into producing heat. The target is a high-melting point, heavy metal (usually tungsten) embedded in a copper rod so that heat is conducted away from the target.

A typical X-ray spectrum is shown in Fig. 10.1. X-ray spectra have two distinct components.

(i)   A background of continuous radiation, the minimum wavelength of which depends on the operating voltage of the tube, i.e. on the energy of the bombarding electrons.

(ii)   Very intense emission at a few discrete wavelengths (an X-ray line spectrum). These wavelengths are characteristic of the target material and are independent of the operating voltage.

The continuous background (or bremsstrahlung radiation) is produced by electrons colliding with the target and being decelerated. The energy of the emitted X-ray quantum is equal to the energy lost in the deceleration. An electron may lose any fraction of its energy in this process. The most energetic X-rays (i.e. those whose wavelength is λ_{min}) are the result of bombarding electrons losing all their energy at once. Since the energy of the electrons depends on the operating voltage, so too does λ_{min} (see Example 10.1). X-rays with longer wavelengths are the result of electrons losing less than their total energy.

The line spectrum is the result of electron transitions within the atoms of the target material. The electrons that bombard the target are very energetic ($\sim 100\,\text{keV}$) and are capable of knocking electrons out of deep-lying energy levels of the target atoms. (This corresponds to removing an electron from an inner orbit on the Bohr model.) If an outer electron then 'falls' into one of these vacancies, an X-ray photon is emitted. The wavelength of the X-ray is given by

(a)

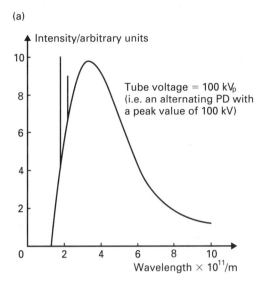

(b)

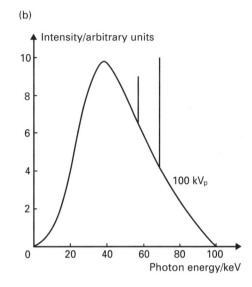

Fig. 10.1
X-ray spectrum of
tungsten as a function
of (a) wavelength, (b)
photon energy

$E = hc/\lambda$, where E is the difference in energy of the levels involved, c is the speed of light and h is Planck's constant. Since the energy levels are characteristic of the target atoms, so too are the wavelengths of the X-rays produced in this way.

Note   The essential difference between transitions that give rise to X-rays and those that produce light is that X-ray transitions involve deep-lying energy levels, optical transitions do not.

EXAMPLE 10.1

Calculate the wavelength of the most energetic X-rays produced by a tube operating at 1.0×10^5 V.

$(h = 6.6 \times 10^{-34}$ J s, $e = 1.6 \times 10^{-19}$ C, $c = 3.0 \times 10^8$ m s^{-1}.)

Solution

The most energetic X-rays are those produced by electrons which lose all their kinetic energy on impact.

$$\text{KE on impact} = \text{work done by accelerating PD}$$
$$= 1.6 \times 10^{-19} \times 1.0 \times 10^5 \quad (\text{by W} = \text{QV})$$
$$= 1.6 \times 10^{-14} \text{ joules}$$

$\therefore$ Maximum KE lost $= 1.6 \times 10^{-14}$ joules

The energy of the corresponding X-ray quantum is hc/λ_{min} and therefore

$$\frac{hc}{\lambda_{min}} = 1.6 \times 10^{-14}$$

i.e. $\lambda_{min} = \dfrac{6.6 \times 10^{-34} \times 3.0 \times 10^8}{1.6 \times 10^{-14}}$

i.e. $\lambda_{min} = 1.24 \times 10^{-11}\,\text{m}$

Note   It should now be clear that the maximum photon energy, E_{max}, and the minimum wavelength, λ_{min}, are given by

$$E_{max} = eV \qquad \text{and} \qquad \lambda_{min} = \dfrac{hc}{eV}$$

where V is the operating PD.

EXAMPLE 10.2

The current in a water-cooled X-ray tube operating at 60 kV is 30 mA. 99% of the energy supplied to the tube is converted into heat at the target and is removed by water flowing at a rate of 0.060 kg s^{-1}. Calculate: (a) the rate at which energy is being supplied to the tube, (b) the increase in temperature of the cooling water. (Specific heat capacity of water $= 4.2 \times 10^3\,\text{J}\,\text{kg}^{-1}\,{}^{\circ}\text{C}^{-1}$.)

Solution

(a) Rate of supply of electrical energy $= 60 \times 10^3 \times 30 \times 10^{-3}$

$$= 1.8 \times 10^3\,\text{J}\,\text{s}^{-1}$$

(b) Rate of production of heat $= 0.99 \times 1.8 \times 10^3$

$$= 1782\,\text{J}\,\text{s}^{-1}$$

Let $\Delta\theta =$ increase in temperature of water

Heat gained by water in 1 s $= 0.060 \times 4.2 \times 10^3\,\Delta\theta$

$$= 252\,\Delta\theta$$

Since all the heat produced is removed by the water,

$$252\,\Delta\theta = 1782$$

$\therefore \qquad \Delta\theta = \dfrac{1782}{252} = 7.1\,{}^{\circ}\text{C}$

QUESTIONS 10A

1. The most energetic X-rays produced by a particular X-ray tube have a wavelength of 2.1×10^{-11} m. What is the operating PD of the tube?

$(e = 1.6 \times 10^{-19}\,\text{C}, h = 6.6 \times 10^{-34}\,\text{J}\,\text{s}, c = 3.0 \times 10^8\,\text{m}\,\text{s}^{-1}.)$

2. What is the wavelength of a beam of X-rays that consists of photons of energy 30 keV? ($1\,\mathrm{eV} = 1.6 \times 10^{-19}\,\mathrm{J}$, $h = 6.6 \times 10^{-34}\,\mathrm{J\,s}$, $c = 3.0 \times 10^{8}\,\mathrm{m\,s^{-1}}$.)

3. An X-ray tube which is 1% efficient produces X-ray energy at a rate of $20\,\mathrm{J\,s^{-1}}$. Calculate the current in the tube if the operating PD is $50\,\mathrm{kV}$.

4.
$$n = 3 \quad\text{——————}\quad -11 \times 10^{3}\,\mathrm{eV}$$
$$n = 2 \quad\text{——————}\quad -26 \times 10^{3}\,\mathrm{eV}$$
$$n = 1 \quad\text{——————}\quad -98 \times 10^{3}\,\mathrm{eV}$$
The diagram shows the three lowest energy levels of an atom of the target material of an X-ray tube. What is the minimum PD at which the tube can operate: **(a)** if the transition $n = 3$ to $n = 1$ is to be possible, **(b)** if $n = 2$ to $n = 1$ is to be possible? **(c)** What is the wavelength corresponding to the transition $n = 3$ to $n = 1$? ($e = 1.6 \times 10^{-19}\,\mathrm{C}$, $h = 6.6 \times 10^{-34}\,\mathrm{J\,s}$, $c = 3.0 \times 10^{8}\,\mathrm{m\,s^{-1}}$.)

10.3 QUALITY AND INTENSITY OF X-RAYS

The quality of an X-ray beam is a term used to describe its penetrating power. Increasing the voltage across an X-ray tube increases the maximum photon energy (Fig. 10.2) and therefore increases the penetrating power (i.e. the quality) of the beam.

Fig. 10.2
The effect of changing tube voltage

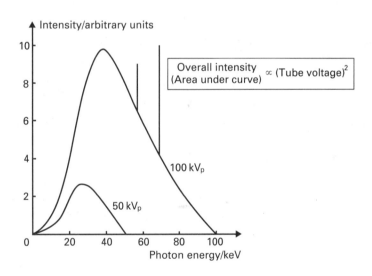

X-rays with low penetrating power are called **soft X-rays**, those with high penetrating power are called **hard X-rays**.

Increasing the PD across an X-ray tube increases the overall intensity of the X-rays (Fig. 10.2) because it increases the energy with which the electrons hit the target and so makes more energy available for X-ray production. Note also that the peak of the intensity distribution shifts to a higher photon energy and that more characteristic lines may be produced.

Increasing the filament current increases the overall intensity of the X-rays (Fig. 10.3) because it increases the number of electrons hitting the target.

Fig. 10.3
The effect of changing
filament current

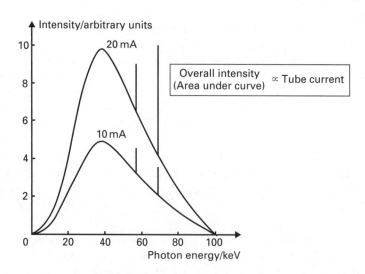

Notes   (i)   To a good approximation:

$$\textbf{Overall intensity} \propto \textbf{(Tube voltage)}^2$$

and

$$\textbf{Overall intensity} \propto \textbf{Tube current}$$

(ii)   The photon energy at which the maximum intensity occurs is about 40% of
the maximum photon energy (see Figs. 10.1(b), 10.2 and 10.3).

QUESTIONS 10B

1. Some X-ray tubes operate on an <u>alternating</u> supply. What differences (if any) would there be in **(a)** the overall intensity, **(b)** the maximum photon energy of the X-rays produced by a tube operating at a steady 100 kV and one operating on an alternating PD with a peak value of 100 kV?

10.4 ROTATING-ANODE X-RAY TUBE

Most diagnostic applications of X-rays require tubes with higher intensity outputs than those used for therapy. In order to achieve this greater output without overheating the target, diagnostic tubes usually have a **rotating anode** (Fig. 10.4).

Fig. 10.4
Rotating anode X-ray
tube

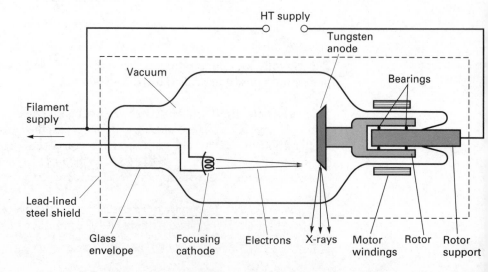

The anode consists of a bevelled tungsten disc mounted on a molybdenum stem. It is made to rotate, by means of an induction motor, at (typically) $3000 \, \text{rev min}^{-1}$. Because of the rotation, the heat loading at any particular point on the target is reduced, thus permitting higher X-ray intensities than those of fixed-target tubes. Cooling is due to radiation rather than conduction.

10.5 ATTENUATION OF X-RAYS

The attenuation of an X-ray beam is the reduction in its intensity due to its passage through matter.

If a beam of <u>monoenergetic</u> (i.e. single wavelength) X-rays of intensity I_0 is incident on an absorber of thickness x (Fig. 10.5), the intensity, I, of the transmitted beam is given by

$$I = I_0 \, e^{-\mu x}$$

[10.1]

where μ = a constant for a given material and given photon energy. It is called the **linear attenuation coefficient** of the material at the energy concerned. (Unit = m^{-1}.)

Fig. 10.5
The effect of an absorber on X-rays

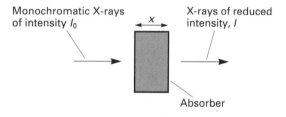

The intensity of a <u>monoenergetic</u> beam decreases exponentially with absorber thickness, i.e. equal thicknesses of absorber absorb equal fractions of the energy regardless of the intensity (Fig. 10.6).

Fig. 10.6
The effect of absorber thickness on X-ray intensity

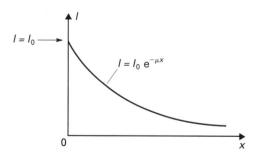

The half-value thickness or HVT ($x_{1/2}$) is the thickness of material that reduces the intensity of an X-ray beam to half its original value.

When $x = x_{1/2}$, $I = I_0/2$ and therefore by equation [10.1]

$$\frac{I_0}{2} = I_0\, e^{-\mu x_{1/2}}$$

$$\therefore \quad \frac{1}{2} = e^{-\mu x_{1/2}}$$

$$\therefore \quad \log_e(1/2) = -\mu x_{1/2}$$

i.e. $$x_{1/2} = \frac{0.6931}{\mu}$$

The penetrating power or quality of different X-ray beams can be compared in terms of their HVTs in a given material – **the greater the HVT, the greater the penetrating power**.

Measurement of HVT

HVTs can be measured by placing different thicknesses (x) of a given material in the path of an X-ray beam and measuring the intensity (I) transmitted at each value of x. A plot of I against x provides the HVT. (The graph will not be exponential for a heterogeneous (i.e. mixed wavelength) beam but the HVT can still be obtained.)

Measurement of μ

For a monoenergetic beam

$$I = I_0\, e^{-\mu x} \quad \text{(equation [10.1])}$$

$$\therefore \quad \log_e I = \log_e (I_0\, e^{-\mu x})$$

$$\therefore \quad \log_e I = \log_e I_0 + \log_e (e^{-\mu x})$$

$$\therefore \quad \log_e I = -\mu x + \log_e I_0$$

Therefore a plot of $\log_e I$ against x (Fig. 10.7) is a straight line with a gradient of $-\mu$, hence μ.

Fig. 10.7
Plot to determine the linear attenuation coefficient

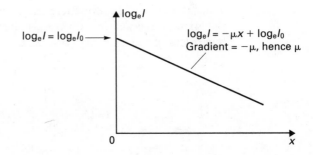

QUESTIONS 10C

1. The HVT of 30 keV X-rays in aluminium is 2.4 mm. If the initial intensity of such a beam is $4.0 \times 10^2\, \text{kW m}^{-2}$, what is its intensity after passing through **(a)** 4.8 mm, **(b)** 9.6 mm of aluminium?

2. **(a)** Calculate the linear attenuation coefficient of aluminium for the X-ray beam in Question 1.

   (b) What is the intensity of the beam after passing through 1.2 mm of aluminium?

Notes (i) The X-ray beams used in medicine are heterogeneous, i.e. they are composed of a whole range of photon energies. **A heterogeneous beam becomes progressively harder (i.e. $x_{1/2}$ increases) as it passes through greater and greater thicknesses of absorber.** This is because any given thickness of absorber absorbs a greater proportion of low-energy photons than high-energy photons and therefore the average photon energy progressively increases.

(ii) The intensity of a beam of X-rays decreases with distance from the source of the beam simply because the beam diverges. This leads to a reduction in intensity in accordance with the **inverse square law** (see section 11.8).

QUESTIONS 10D

1. An X-ray beam comprising 20 keV and 30 keV photons in equal proportions is incident on an absorber for which the HVTs of the two components are 2.0 mm and 3.0 mm respectively. What is the ratio of 30 keV photons to 20 keV photons after the beam has passed through **(a)** 6.0 mm, **(b)** 12 mm of the absorber? What happens to the overall HVT of the beam as it passes through greater and greater thicknesses of absorber?

Mass Attenuation Coefficient (μ_m)

A quantity called the **mass attenuation coefficient** (μ_m) is sometimes used. It is related to the linear attenuation coefficient (μ) of a material by

$$\mu_m = \frac{\mu}{\rho}$$

where

$\rho = $ density of the material

For any given substance the value of μ_m (unlike the value of μ) is independent of the density of the substance. For example, the mass attenuation coefficient of water is $0.017 \text{ m}^2 \text{ kg}^{-1}$ regardless of whether the water is in the form of ice, liquid water or water vapour.

10.6 INTERACTION OF X-RAYS WITH MATTER

Four of the processes by which X-rays (and γ-rays) interact with matter are described below.

Simple (or Elastic) Scattering

This occurs when the energy of the X-ray photon is less than that required to remove an electron from an atom of the material through which it is passing. The photon interacts with an orbiting electron in such a way that it is deflected from its path without losing any of its energy. In this process, therefore, the material scatters the X-ray beam but absorbs none of its energy.

Photoelectric Absorption

In this process (Fig. 10.8) the incident X-ray photon gives up the whole of it
energy to an orbiting electron and ejects it from its atom. The electron is known as
a **photoelectron**. An outer electron then 'falls' into the vacancy created, releasing
radiation that is characteristic of the absorber atom. When heavy atoms are
involved, this characteristic radiation may be in the form of X-rays (of lower
energy than those which are absorbed). When the absorber is bone or body tissue
only light atoms are involved and the characteristic radiation is of such low energy
that it is immediately absorbed in the material itself. The photoelectron dissipate
its energy by ionizing atoms along its path.

Fig. 10.8
Photoelectric absorption

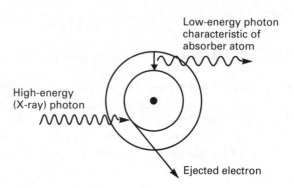

Compton Scattering

In Compton scattering (Fig. 10.9) the X-ray photon transfers some of its energy to
an orbiting electron. The electron, known as a **recoil electron**, is ejected from its
atom and a photon of lower energy (i.e. longer wavelength) than the original
photon goes off in a different direction. The greater the angle through which the
photon is scattered, the greater its loss of energy. The recoil electron dissipates its
energy through ionization.

Fig. 10.9
Compton scattering

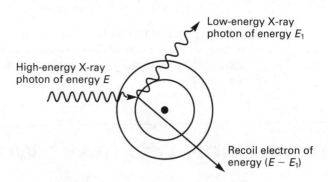

Pair Production

An X-ray photon with energy in excess of 1.022 MeV can interact with the nucleus
of an atom in such a way that the photon ceases to exist and an electron–positron
pair is created in its place*. The electron dissipates its energy through ionization.
The positron is annihilated by an electron, creating two identical photons that
move in opposite directions in order to conserve momentum.

* The total rest mass of an electron and a positron corresponds to an energy of 1.022 MeV.

10.7 RELATIVE IMPORTANCE OF ATTENUATION MECHANISMS

The relative importance of the various attenuation processes depends on both the atomic number of the absorber (Z) and the photon energy (E) – see Table 10.1.

Table 10.1
The relative importance of the various attenuation processes

Mechanism	Dependence of μ_m on E	Dependence of μ_m on Z	Comment
Simple scatter	Decreases as E increases	Increases as Z increases	Significant only at photon energies below 20 keV
Photoelectric absorption	Decreases <u>markedly</u> as E increases ($\mu_m \propto 1/E^3$)	Increases <u>markedly</u> as Z increases ($\mu_m \propto Z^3$)	The dominant process in tubes used for diagnosis, i.e. at 30 keV
Compton scatter	Decreases slightly as E increases	Independent of Z	The dominant process in tubes used for therapy, i.e. at 0.5 to 5 MeV
Pair production	Increases as E increases	Increases as Z increases	Occurs only above 1.022 MeV and is significant only above 5 MeV

The peak intensity of X-ray tubes used for <u>diagnosis</u> occurs at a photon energy of about 30 keV. Photoelectric absorption is the dominant attenuation mechanism at this energy. **Photoelectric absorption depends on Z^3** and therefore bone (for which $Z \approx 14$) absorbs the radiation much more strongly than soft tissue (for which $Z \approx 7$). This produces strong contrast between the X-ray images of bone and soft tissue. The greater density of bone increases the contrast even more.

Tubes used for <u>therapy</u> usually operate at much higher energies (0.5 to 5 MeV) where the dominant attenuation mechanism is Compton scattering. **Compton scattering is independent of Z** and therefore there is no preferential absorption by bone (which would be potentially damaging and would serve no purpose).

Note Photoelectric absorption is the dominant X-ray attenuation mechanism at photon energies up to 30 keV (approximately) in soft tissue and up to 50 keV in bone. Compton scatter dominates between 30 keV and 22 MeV in soft tissue and between 50 keV and 18 MeV in bone. Pair production is the dominant mechanism above 22 MeV in soft tissue and above 18 MeV in bone.

10.8 FILTRATION

Except for the special case of superficial (i.e. surface) radiotherapy, the X-rays used in radiotherapy treatment need to penetrate deep inside the patient's body; those used for diagnosis need to pass right through it.

The radiation emitted by an X-ray tube is **heterogeneous** i.e. it is made up of photons with a range of different energies. The low-energy photons are usually filtered out of the beam by passing it through a suitable filter. This is because most of these photons would merely be unproductively absorbed in the patient's skin and surface tissue; only very few of them would reach the deeper parts of the body or emerge from the far side of it, and so would add to the radiation dose received by the patient without serving any useful purpose.

The filter must remove low-energy photons in preference to high-energy photons and therefore must be made of a material in which photoelectric absorption, which varies as $1/(\text{photon energy})^3$, dominates the other attenuation processes in the energy range concerned. Since the probability of photoelectric absorption is proportional to Z^3, high-Z materials are used. A second advantage of making use of photoelectric absorption is that it absorbs the low-energy photons and therefore removes them completely; Compton scattering would simply replace them with others (of lower energy) travelling in different directions.

The filter in tubes used for diagnosis is usually a few millimetres of aluminium. Tubes designed for therapy usually operate at higher voltages than diagnostic tubes. Aluminium removes too high a proportion of the photons by Compton scattering (which barely distinguishes between high and low-energy photons) at these higher voltages and therefore higher-Z materials such as copper, tin, lead and gold are used.

The effect of a filter is shown in Fig. 10.10. Note that there is a large reduction in the overall intensity because the unfiltered beam contains a high proportion of low-energy photons. Because the filter increases the proportion of high-energy photons, it follows that it increases the HVT of the beam.

Fig. 10.10
The effect of filtration on a typical X-ray spectrum

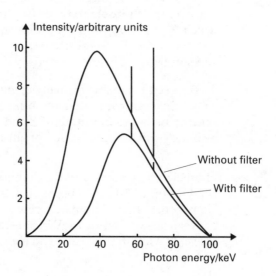

10.9 BEAM SIZE AND ALIGNMENT

The cross-section of an X-ray beam and its alignment can be controlled by the use of a device called a **light-beam diaphragm** (**LBD**). It consists of two pairs of lead sheets (Fig. 10.11(a)) which can be moved at right angles to each other. They determine the cross-section of the beam by absorbing the radiation that does not pass through the rectangular opening at their centre.

The device incorporates a lamp and mirror arrangement (Fig. 10.11(b)) which produces a light beam that has the same cross-section and the same direction as the X-ray beam, and which therefore allows the operator to see (literally) the area that will be exposed to X-rays.

Fig. 10.11
The light-beam
diaphragm (a) beam
definer (b) alignment
device

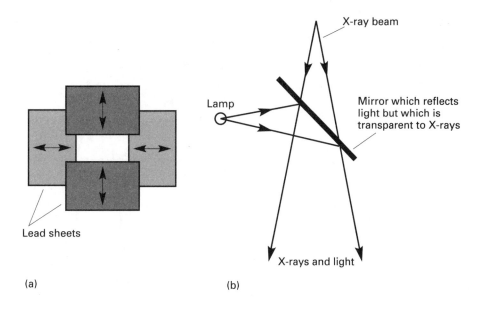

(a)   (b)

10.10 QUALITY OF X-RAY IMAGE

Some of the factors that affect the quality of an X-ray image are discussed below.

Size of Focal Spot

X-rays cannot be focused and therefore an X-ray image of a bone, say, is simply
a shadow of the bone. Because X-rays originate from a small (rectangular) area of
the target known as the **focal spot**, rather than from a point, the images (shadows)
they produce do not have sharp edges (see Fig. 10.12). The focal spot can be made
smaller (and therefore the image made sharper) by using a more focused electron
beam. The extent to which the electrons can be focused without generating
excessive heat in the target is determined by the orientation of the target (see
Fig. 10.13). The orientation is such that the X-rays effectively originate from a
smaller area than that from which they actually originate.

It should be clear from studying Fig. 10.12(b) that the sharpness of the image can
be increased by putting the film closer to the object (i.e. by decreasing the **film-to-**

Fig. 10.12
To illustrate the effect of
the size of the focus on
image sharpness

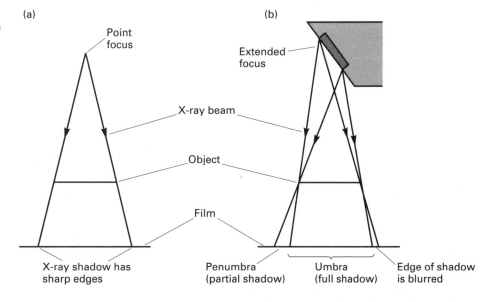

Fig. 10.13
Effect of target
orientation on the size of
the focal spot

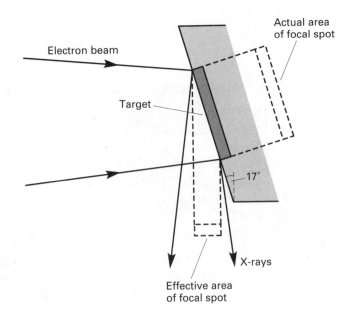

object distance). It should also be clear that the sharpness can be increased by increasing the **focus-to-film distance** (without altering the film-to-object distance). Because of the inverse square law effect (section 11.8), increasing the focus-to-film distance decreases the intensity of the X-rays and therefore requires the use of a longer exposure time.

Scattering

The adverse effect of scattered radiation, which would decrease both the sharpness and the contrast of the image, can be almost totally eliminated by placing a **grid** between the patient and the film (Fig. 10.14). The grid consists of a large number of strips of lead foil separated from each other by a material which is transparent to

Fig. 10.14
Anti-scatter X-ray grid

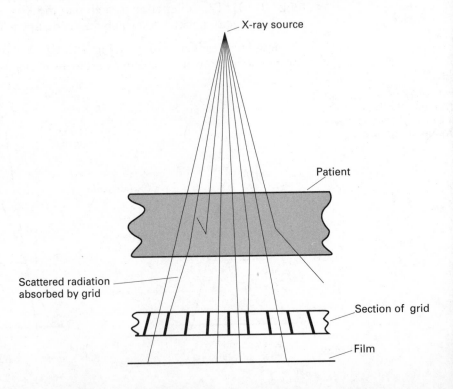

X-rays. The strips, which are about 5 mm deep and 0.05 mm thick, are about 0.5 mm apart and stand on edge pointing towards the X-ray source. Most of the primary (unscattered) radiation passes through the gaps between the lead strips and reaches the film. The scattered radiation is absorbed by the lead and is therefore prevented from reaching the film.

The grid is often made to move from side to side across the film in order to blur the shadow cast by the strips, which would otherwise degrade the image.

Movement Blur

Examinations of organs that are subject to involuntary movements (e.g. the stomach and the bowel) require short exposure times in order to avoid blurring of the X-ray image.

10.11 CONTRAST MEDIA

X-ray images are clearest when there are large differences in proton number and/or density between the part of the body under investigation and those around it. When this is not the case, the contrast, and therefore the clarity, of the image can be increased by administering an artificial **contrast medium** to the patient. These are usually compounds of high-proton number elements such as barium and iodine. A common example is the so-called **barium meal**. The patient swallows a thick suspension of barium sulphate which passes into the gastrointestinal tract. The barium attenuates the X-ray beam more strongly than the surrounding tissue and therefore the tract shows up clearly against the background.

X-ray image of the large intestine after a barium enema

10.12 INTENSIFYING SCREENS

Photographic film is relatively insensitive to X-rays and therefore **intensifying screens** are used in order to avoid the long exposure times that would otherwise be required. This keeps the radiation dose received by the patient down to an acceptable level.

The so-called **intensifying screen cassette** (Fig. 10.15) consists of a sheet of double-sided film sandwiched between, and in intimate contact with, two intensifying screens. The screens are sheets of <u>white</u> plastic onto which fluorescent crystals have been bonded. A metal plate at the back of the cassette prevents radiation being scattered back onto the film from the couch beneath it.

Fig. 10.15
Intensifying cassette

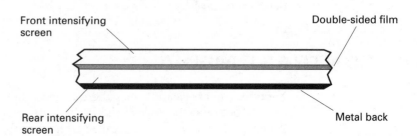

The fluorescent crystals absorb X-rays which have passed through the patient and re-emit the X-ray energy as visible light. Some of this light heads directly towards the film, and most of that which would otherwise be 'lost' is reflected back to it by the white plastic. Since the film is more sensitive to light than to X-rays, there is a significant decrease in exposure time – by a factor of as much as 250 with the most recent types. The crystals emit light in all directions and it is therefore vital that they are as close as possible to the film for there would otherwise be an unacceptable loss of resolution.

10.13 FLUOROSCOPY

In **X-ray fluoroscopy**, X-rays are passed through the patient and onto a fluorescent screen to produce an immediately visible image. This has the advantage over photographic film in that it allows <u>dynamic</u> processes such as blood flow to be investigated. Unfortunately, unacceptably high X-ray intensities would be needed to produce images that could be viewed directly. A device known as an **image intensifier** can increase the brightness by a factor of over a thousand and allows the radiation dose to be cut by up to 90% of the unintensified level. The intensifier (Fig. 10.16) has a fluorescent screen in contact with a photocathode. This combination converts the X-rays first to visible photons and then to electrons. The number of electrons at any point on the photocathode is directly proportional to the X-ray intensity at that point, and hence to the intensity transmitted by the patient.

The electrons produced by the photocathode are then accelerated through a PD of about 20 kV, using a series of focusing anodes, towards a second fluorescent screen. The increased energy and concentration of the electrons creates an image that is very much brighter than that on the first screen, and which is usually picked up by a TV camera and fed to a TV monitor or video recorder.

Fig. 10.16
Image intensifier

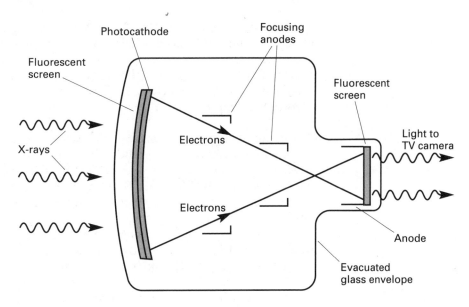

Fluoroscopy is a technique that is used sparingly, for despite image intensification, the radiation dose to the patient is still significantly higher than that in a standard radiographic examination. Dose 'savings' can be made by using short bursts of X-rays rather than a truly continuous exposure, but even so the examinations remain relatively dose-intensive.

10.14 COMPUTED TOMOGRAPHY (CT)

Computed tomography (**CT**) is a means of producing a two-dimensional image of a slice of the body. A fan-shaped beam of X-rays (Fig. 10.17) passes through the slice being imaged to an array of about two thousand tiny radiation detectors

Fig. 10.17
CT scanner

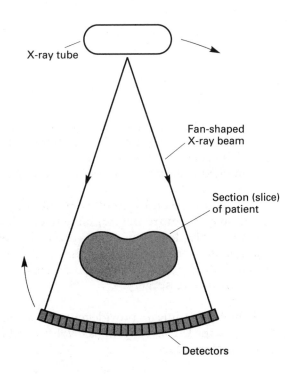

arranged in a circular arc. The X-ray source and the detector array are rotated about the centre of the slice. Whilst this is happening, a few hundred X-ray pulses are passed through the patient, each from a different position of the source. The transmitted intensity picked up by each detector from each position of the source is relayed to a computer which, at the completion of the scan, uses the information to construct the image.

The computer models the image plane as a set of tiny cells. Its task is to determine the attenuation that each cell produces. (It then constructs an image in which the brightness at any point is determined by the attenuation produced in the corresponding cell.) We can illustrate the principle by means of Fig. 10.18 in which the image plane is represented by six cells, producing (unknown) attenuations a, b, c, d, e and f. If a beam of X-rays is incident from the left (Fig. 10.18(a)), the transmitted beams give information concerning the total attenuation produced in each of the three rows, and this can be expressed as three equations. Changing the orientation by $45°$ (Fig. 10.18(b)) gives four more equations. We now have more than enough equations to find the values of a, b, c, d, e and f. We leave it as an exercise for the reader to show that $a = 2$, $b = 5$, $c = 7$, $d = 3$, $e = 2$ and $f = 4$.

Fig. 10.18
To illustrate CT image construction

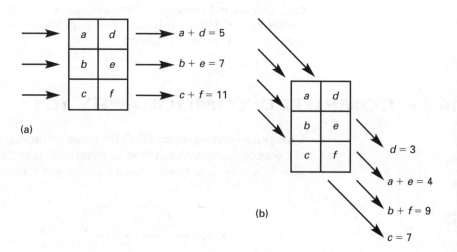

We have illustrated the technique with just 6 cells – the actual process involves hundreds of thousands of calculations and requires a powerful computer with sophisticated software. The image can be displayed on a TV screen or stored and combined with images of parallel slices to produce a three-dimensional view.

CT scanners can detect very slight differences in X-ray attenuation and are therefore particularly useful for examining soft tissue. They are increasingly being used to produce detailed images of the brain and abdominal organs, such as the liver and the kidneys, with resolutions of better than 1 mm. The three-dimensional images are useful for planning reconstructive surgery.

Although CT images provide more detailed information than conventional X-ray images, the CT investigations involve significantly higher radiation doses and are much more expensive.

Three dimensional
computed tomography
(CT) scan of the brain.
(Scan data of the left
cerebral hemisphere have
been excluded.)

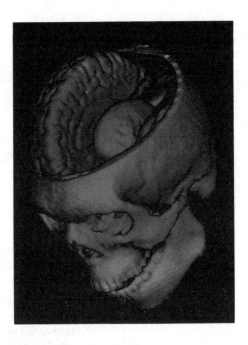

10.15 RADIOTHERAPY USING X-RAYS

The term **radiotherapy** refers to the treatment of a medical condition (usually cancer) by means of X-rays, γ-rays or beams of energetic electrons. X-ray radiotherapy falls into two main classes – **superficial therapy** and **megavoltage (or MV) therapy**. (The so-called 'deep-therapy' treatment is rarely used nowadays and the term is largely obsolete.)

Superficial therapy is used to treat conditions of the skin and surface tissue. The tube voltages employed are such that the X-rays have only low penetrating power and therefore cause little damage to the healthy tissue beneath the area being treated. (The peak X-ray intensity is typically 20 keV.)

Megavoltage therapy is used to treat conditions deep inside the body and has almost completely replaced the lower-voltage techniques once used for this purpose. The electrons used to create the X-ray beam are accelerated to the enormous energies required in a linear accelerator (LINAC). (The peak X-ray intensity is typically 0.5 to 5 MeV.)

Advantages of MV Therapy

(i)   It decreases the damage sustained by the patient's skin. The beam is so penetrating that hardly any of its energy is absorbed by the skin and surface tissue.

(ii)   It reduces damage to bone (see section 10.7).

Rotating Beams

The purpose of radiotherapy is to destroy malignant (i.e. cancer) cells whilst doing as little damage as possible to the healthy tissue and bone around them. One way of achieving this is to aim the beam at the tumour from a number of different directions, i.e. to rotate it about the tumour. This technique is known as **multiple-beam** or **rotating-beam** therapy and produces a considerable cumulative effect at the tumour but a much reduced effect everywhere else.

10.16 TREATMENT PLANNING

Any amount of radiation is potentially harmful to the person being exposed to it and therefore it is important that the following considerations are taken into account.

(i) The likely benefits of exposure to radiation must outweigh the risks involved.

(ii) The radiation dose must be the <u>minimum</u> consistent with obtaining good quality images or destroying malignant cells.

(iii) It must not be possible to obtain equally useful information by less risky methods.

(iv) The beam must be collimated so that only that part of the body that needs to be exposed is exposed.

(v) Increasing the PD across an X-ray tube increases the penetrating power of the X-rays produced but it decreases the proportion of the beam that is attenuated by <u>photoelectric</u> absorption, and therefore decreases the contrast of the X-ray image (see section 10.7). The PD should be high enough to produce the required degree of penetration but not so high that there is insufficient contrast. Another consequence of increasing the tube voltage is that it increases the energy, and therefore the penetrating power, of the <u>scattered</u> radiation. This increases the likelihood of the patient receiving a significant radiation dose in parts of the body some distance away from the part being exposed intentionally. It also increases the chance of scattered radiation escaping from the patient – a potential hazard to hospital staff in the vicinity.

(vi) The radiographer must employ good techniques so that repeat exposures are not required. (This is an important consideration when assessing just what <u>is</u> the minimum dose – an underestimation would necessitate a second exposure and therefore an overall increase in dose.)

CONSOLIDATION

X-rays are high-photon energy (i.e. short-wavelength) electromagnetic radiation.

The intensity of an X-ray beam can be increased by increasing the filament current and/or by increasing the PD across the tube.

The quality of an X-ray beam can be increased by increasing the PD across the tube.

The maximum photon energy, E_{max}, and minimum wavelength, λ_{min} are given by

$$E_{max} = eV \qquad \lambda_{min} = \frac{hc}{eV}$$

where V is the operating PD.

For a <u>monoenergetic</u> (i.e. single wavelength) beam of X-rays

$$I = I_0\, e^{-\mu x}$$

The half-value thickness (HVT) is the thickness of material that halves the intensity of an X-ray beam.

The HVT of a <u>homogeneous</u> (i.e. single wavelength) beam has a fixed value in any given material. The HVT of a <u>heterogeneous</u> (i.e. multiple wavelength) beam increases as it passes through increasing thicknesses of absorber because the beam becomes progressively harder.

Mass attenuation coefficient (μ_m) and linear attenuation coefficient (μ) are related by

$$\mu_\mathrm{m} = \frac{\mu}{\rho}$$

Photoelectric absorption depends on Z^3 and is the dominant attenuation mechanism in X-ray tubes used for diagnosis.

Compton scatter is independent of Z and is the dominant attenuation mechanism in X-ray tubes used for therapy.

The inverse square law for X-rays. The intensity, I, at a distance, d, from a point source of X-rays is given by

$$I \propto \frac{1}{d^2} \quad \text{i.e.} \quad I = \frac{\text{constant}}{d^2}$$

X-ray beams are filtered to remove low-energy photons because these would be absorbed in the patient's skin and surface tissue, adding to the radiation dose without serving any useful purpose. High-Z materials are used as filters in order to make use of photoelectric absorption.

QUESTIONS ON CHAPTER 10

1. An X-ray tube with a tungsten target is operated at a voltage of 50 kV and current 10 mA. The X-ray spectrum from such a tube is shown in the diagram below.

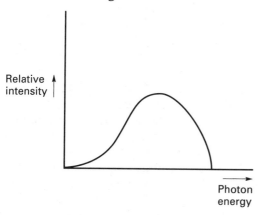

Copy the diagram and add the intensity/energy graphs you would expect to obtain if
(a) the tube were operated at 100 kV and 10 mA,
(b) the tube were operated at 50 kV and 20 mA.
[Label each of your graphs clearly (a) or (b).]
Describe the mechanism known as the *photoelectric effect* by which low energy (30 keV) X-ray photons are attenuated. How does this attenuation depend on the proton number of the absorber? Explain why low energy X-rays are preferred for radiography.

[L, '93]

2. Describe briefly the physical processes by which a beam of X-rays for medical diagnosis is attenuated in passing through a patient's body.
A diagnostic radiography is taken of part of a patient's body containing bones, tissue and air spaces. Describe and account for the appearance of these regions on an X-ray film.

[L, '87]

3. What is meant by the half-value thickness (HVT) of a beam of X-radiation?
A homogeneous X-ray beam has an intensity of 20 MW m^{-2} at a distance of 0.1 m from the source of the X-rays. What will be the intensity
(a) 1.0 m from the source with no filtration,
(b) 1.0 m from the source if the X-rays are passed through an aluminium filter of thickness 6.0 mm? (The HVT for this beam in aluminium is 1.5 mm.)
Explain, without further calculation, how your answers to (a) and (b) above might differ if the original beam was not homogeneous.

[L, '95]

4. (a) (i) Draw a diagram of a rotating anode X-ray tube.
(ii) Label the important features of the tube.
(iii) Why is as much air as possible removed from inside the tube?

(iv) Explain why a rotating anode is used instead of a stationary one.

(v) Why does the anode have a large mass?

(b) In diagnostic X-radiography the beam leaving the X-ray generator usually passes through a thin sheet of aluminium.

(i) Explain why this sheet is used.

(ii) 200 keV X-ray photons are incident normally on an aluminium sheet which has a mass attenuation coefficient of $0.012 \, \mathrm{m^2 \, kg^{-1}}$. If the density of aluminium is $2700 \, \mathrm{kg \, m^{-3}}$ find the linear attenuation coefficient.

(iii) Calculate the thickness of aluminium required to reduce the intensity of the same X-ray beam to 1% of the original intensity.

(c) An important feature of the image produced on an X-ray film is the *contrast*.

(i) Explain what is meant by *contrast*.

(ii) State **one** factor relating to the X-ray beam on which the contrast depends.

(iii) Name **one** device which could be used to increase the contrast on the film. [N, '93]

5. The intensity, I, of a parallel non-homogeneous beam of X-rays is measured after passing through different thicknesses, x, of absorbing material. For $x = 0$, $I = I_0$, and a graph of $\ln(I/I_0)$ against x is plotted as shown below.

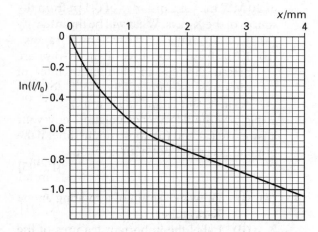

(a) Over what range of x would the homogeneous beam relationship, $I = I_0 \, e^{-\mu x}$ hold? Calculate μ in this range.

(b) Describe and explain what happens to the quality (the spectrum of X-ray wavelengths) of the transmitted X-ray beam as the thickness of the absorber increases from 0 to 3 mm.
Name a material which is used for filters in X-ray diagnosis. [L, '94]

6. (a) X-rays can be used to produce an image on a film of a part of the body.

(i) How does the formation of an X-ray image differ from that of an ultrasonic image?

(ii) Why would the use of ultrasound be preferred to the use of X-rays when imaging a foetus?

(iii) In a radiographic X-ray machine, the measured intensity 0.10 m from the point source of X-rays is $20 \, \mathrm{mW \, m^{-2}}$. What is the intensity 2.0 m from the source?

(b) Name **one** other method (not based on ultrasound or X-rays) which could be used to image the interior of the body. [N, '94]

7. (a) Describe, using a diagram, a method of producing X-rays.

(b) State an advantage of producing X-rays from a small area of the target electrode when an X-ray photograph is required.

(c) What problem can arise as a result of using X-rays for taking photographs? How is the problem minimised both for the radiographer and the patient?

(d) To examine internal structures within the body, ultrasound in the MHz frequency range can be used in order to overcome problems such as the one referred to in (c).

(i) Why is very high frequency ultrasound needed?

(ii) Explain the principles of one method of producing ultrasound. [C, '91]

8. State, without explanation, **three** methods which are used for minimising the absorbed dose to patients during diagnostic radiography.
Calculate the ratio of the transmitted to the incident intensities of an X-ray beam travelling through a layer of aluminium 2 mm thick, the half-value thickness being 3 mm. [N, '88]

9. **(a)** Name and describe the two main processes by which the beam of X-rays from a tube used for diagnostic radiography is attenuated by a thin slab of absorbing material.

(b) Discuss how the magnitude of each process depends on the energy of the X-rays and the nature of the absorber. Hence explain why the optimum photon energy for radiography is around 30 keV.

(c) Explain why it is common to pass the X-ray beam produced by the tube for radiography through a thin sheet of aluminium before irradiation of the patient.   [O & C, '92]

10. **(a) (i)** Sketch a diagram of a typical diagnostic X-ray tube, identifying appropriate parts of the sketch.

(ii) Draw a typical X-ray spectrum generated by such a tube and explain how the intensity of the X-rays may be varied.

(b) (i) Define the *half-value thickness* of a material used as an X-ray absorber.

(ii) Explain why, in producing a radiograph, it is usual to filter the X-ray beam, and suggest a suitable filter material.

(iii) Identify the process chiefly responsible for the absorption of X-rays by the filter material used in radiography.   [N, '82]

11. The diagram below shows the relative intensity, I, of an X-ray beam as a function of X-ray photon energy, E, after the beam has passed through an aluminium filter. For one curve the filter thickness was 3.0 mm, while for the other curve the thickness was 1.0 mm.

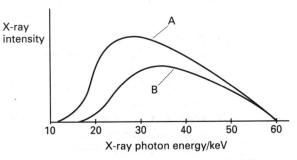

(a) Does curve A or curve B correspond to that for the 3.0 mm aluminium filter? Explain your answer by describing the essential differences between curve A and curve B.

(b) Why do both curves have the same maximum value of E?

(c) Why is the correct filtration of an X-ray beam of importance in diagnostic radiography?

(d) The intensity of a monochromatic beam of X-rays is reduced to $\frac{1}{8}$ of its initial incident value after passing through 3.0 mm of copper. What is the half-value thickness of copper for this radiation?   [L, '88]

12. **(a)** Many diagnostic X-ray machines produce radiation with a photon energy of about 30 keV. The dominant process, at this energy, by which the X-ray beam is attenuated is the photoelectric effect. State what factors control the amount of this attenuation as the beam passes through the human body containing different types of tissue. Show the way in which each factor leads to variations in the intensity of X-rays transmitted.

(b) Explain what are meant by the terms *sharpness* and *contrast* when used to indicate the quality of a medical X-ray image, and discuss one factor which limits each. Describe briefly and explain the use of **two** means by which the quality of an X-ray image can be improved.   [O & C, '93]

13. List **three** factors which affect the quality of a radiographic image produced by X-rays. Explain how, instead of allowing the radiation to form an image directly on suitable film, it is possible to reduce the absorbed dose to the patient and still obtain a photographic image of the same density.   [N, '85]

14. **(a) (i)** Name **three** factors which affect the sharpness of images produced by diagnostic X-rays.

(ii) Name **two** factors which affect the contrast of the X-ray image.

(b) State how information obtainable from a diagnostic X-ray differs from that in an image produced by a gamma camera.   [N, '91]

15. **(a)** With the aid of a labelled diagram, give a description of the way in which X-rays are produced.

(b) How can
   (i) the sharpness,
   (ii) the contrast,
   be controlled in X-ray imaging?
(c) How can the harmful effects of X-rays be minimised when making X-ray images of patients? [C, '95]

16. **(a)** An X-ray tube operating at 65 kV has a tube current of 120 mA. It produces an X-ray beam with an efficiency of 0.8%. Calculate the intensity of the conical X-ray beam where it has a radius of 20 cm.
   (b) Describe, with the aid of a diagram, the structure and mode of use of a grid which can be used to reduce the amount of scattered radiation reaching the film in diagnostic radiology. [L, '92]

17. As a beam of X-radiation passes through tissues, some of the photons are scattered. This scatter has three important effects:
   —it contributes to the radiation dose received by the patient,
   —it contributes to the radiation dose received by the medical staff, and
   —it degrades the quality of the radiographic image.
   Explain the attenuation mechanisms called **(i)** Compton scatter and **(ii)** simple scatter. Above what photon energy range is Compton scatter the dominant scattering process in soft tissue?
   The diagrams show the effect of Compton scatter on two patients, P and Q, subject to incident X-ray beams of different energy from sources A and B. The short arrowed lines show the path of scattered photons.

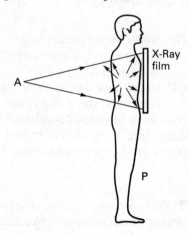

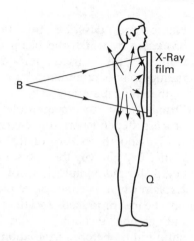

(a) Explain which X-ray source has the larger tube voltage.
(b) Suggest precautions which might be taken to reduce the dose received by the medical staff.
(c) What is usually done to prevent the scattered photons from degrading the image quality? Illustrate your answer with a simple diagram. [L, '92]

18. **(a)** Describe the *photoelectric* and *Compton effects*, explaining their importance in the attenuation of X-rays by matter.
   Explain, in the case of each process, why, when an X-ray is absorbed, electromagnetic radiation of larger wavelength than that of the original X-ray, is produced.
   (b) Contrast the usefulness of Geiger counters and scintillation counters for the detection of X-rays.
   (c) The intensity of a particular X-ray beam is decreased by 35% when it passes through an aluminium sheet 2 mm thick.
   Estimate the total thickness of aluminium which would decrease the intensity of the same original beam by 80%.
   [O & C, '91]

19. **(a)** **(i)** An intensifying screen is often used in producing images of parts of the body by X-rays. Describe the structure of such a screen and explain why it is used.
      (ii) Give **two** reasons why an image intensifier might be preferred for observing such images.
   (b) Describe, with the aid of a labelled diagram, the construction and mode of action of an *X-ray image intensifier*.
   [N, '84]

11
RADIOISOTOPES

11.1 THE NUCLEUS

Every atom has a central, positively charged nucleus. Nuclear diameters are $\sim 10^{-15}$ m, atomic diameters are $\sim 10^{-10}$ m. Over 99.9% of the mass of an atom is in its nucleus. Atomic nuclei are unaffected by chemical reactions.

Nuclei contain **protons** and **neutrons*** which, because they are constituents of nuclei, are collectively referred to as **nucleons**. Their properties are compared with those of the electron in Table 11.1. The charge on the proton is equal and opposite to that on the electron, and it follows that (neutral) atoms contain equal numbers of protons and electrons.

Table 11.1
The nucleons compared with the electron

	Electron	Proton	Neutron
Mass	$m_e = 9.110 \times 10^{-31}$ kg	$m_p = 1836\, m_e$	$m_n = 1839\, m_e$
Charge	-1.602×10^{-19} C	$+1.602 \times 10^{-19}$ C	Zero

11.2 ISOTOPES

Two atoms which have the same number of protons but different numbers of neutrons are said to be **isotopes** of each other. It follows that each atom contains the same number of electrons as the other and, therefore, that their chemical properties are identical. Isotopes cannot be separated by chemical means. Some elements have only one naturally occurring isotope (gold and cobalt are examples); tin has the largest number – ten. Isotopes that are radioactive (section 11.3) are known as **radioisotopes**.

> **The atomic number (or proton number)** Z of an element is the number of protons in the nucleus of an atom of the element.
>
> **The mass number (or nucleon number)** A of an atom is the number of nucleons (i.e. protons + neutrons) in its nucleus.

The various isotopes of an element whose chemical symbol is represented by X are distinguished by using a symbol of the form

$$^{A}_{Z}\text{X}$$

where A and Z are respectively the mass number and atomic number of the

---

* The nucleus of the common isotope of hydrogen is an exception – it has a single proton and no neutron.

isotope. The most abundant isotope of lithium (lithium 7) has 3 protons and 4 neutrons, i.e. $Z = 3$ and $A = 3 + 4 = 7$, and therefore it is represented by ^7_3Li Lithium 6 has only 3 neutrons and is written as ^6_3Li. The Z value is sometimes omitted because it gives the same information as the chemical symbol (for example, all lithium atoms have 3 protons).

Hydrogen is exceptional in that its three isotopes are given different names. The most abundant isotope has one proton and no neutron and is actually called hydrogen (^1_1H). The other isotopes are **deuterium** (^2_1D or ^2_1H) and **tritium** (^3_1T or ^3_1H). A deuterium nucleus is called a **deuteron**.

The term **nuclide** is used to specify an atom with a particular number of protons and a particular number of neutrons. Thus ^6_3Li, ^7_3Li, $^{16}_8\text{O}$ and $^{18}_8\text{O}$ are four different nuclides.

QUESTIONS 11A

1. How many of each of the following particles are there in a single atom of iron 56 ($^{56}_{26}\text{Fe}$)?
   (a) electrons, (b) protons, (c) neutrons, (d) nucleons, (e) negatively charged particles, (f) positively charged particles, (g) neutral particles.

11.3 RADIOACTIVE DECAY

In 1896 Henri Becquerel noticed that some photographic plates that had been stored close to a uranium compound had become fogged. He showed that the fogging was due to "radiations"* emitted by the uranium. The phenomenon is called **radioactivity** or **radioactive decay** and the 'radiations' are emitted when an unstable nucleus disintegrates to acquire a more stable state. The **disintegration is spontaneous** and most commonly involves the emission of an α-**particle** or a β-**particle**. In both α-emission and β-emission the **parent nucleus** (i.e. the emitting nucleus) undergoes a change of atomic number and therefore becomes the nucleus of a different element. The new nucleus is called the **daughter nucleus** or the **decay product**. The daughter nucleus is often in an excited state when it is formed, in which case it reaches its ground state by emitting a γ-**ray**. Though most nuclides emit either α-particles or β-particles, some emit both. For example 64% of $^{212}_{83}\text{Bi}$ nuclei emit β-particles and 36% emit α-particles.

Henri Becquerel

---

* Many of the 'radiations' are in fact particles; the term was applied before this was realized and is still in use.

11.4 THE EXPONENTIAL LAW OF RADIOACTIVE DECAY

Radioactive nuclei disintegrate spontaneously; the process cannot be speeded up or slowed down. It follows that **for large numbers of any particular species of nuclei the rate of decay is proportional to the number of parent nuclei present**. If there are N parent nuclei present at time t, the rate of increase of N is dN/dt and therefore the rate of decrease, i.e. the rate of decay, is $-dN/dt$. It follows that

$$-\frac{dN}{dt} = \lambda N$$

or $\qquad \dfrac{dN}{dt} = -\lambda N$ $\hfill$ [11.1]

where

$\qquad \lambda =$ a (positive) constant of proportionality called the **decay constant**. (Unit $= s^{-1}$.)

$\qquad -dN/dt =$ the rate of decay and is called the **activity** of the source. When used in equation [11.1] the activity must be expressed in the relevant SI unit – the becquerel. One **becquerel** (Bq) is equal to an activity of one disintegration per second.

Notes (i) Rearranging equation [11.1] gives

$$\lambda = \frac{-dN/dt}{N}$$

i.e. $\quad \lambda = \dfrac{\text{Number of nuclei decaying per unit time}}{\text{Number of parent nuclei present}}$

i.e. $\quad \lambda =$ Fraction of nuclei decaying per unit time

(ii) Activity used to be expressed in curies. One **curie** (Ci) is defined as (exactly) 3.7×10^{10} disintegrations per second.

Solving equation [11.1] gives

$$N = N_0 e^{-\lambda t}$$ $\hfill$ [11.2]

Equation [11.2] expresses the exponential nature of radioactive decay, i.e. that the number of nuclei <u>remaining</u> after time t (i.e. the number of parent nuclei) decreases exponentially with time (see Fig. 11.1). It is known as the **exponential law of radioactive decay**.

Fig. 11.1
Graph to illustrate the exponential nature of radioactive decay

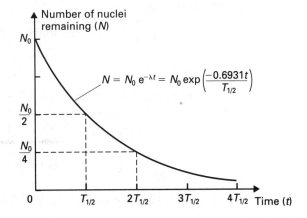

Thus, although it is impossible to predict when any particular nucleus w~~ disintegrate, it is possible to say what proportion of a large number of nuclei w~~ disintegrate in any given time.

11.5 HALF-LIFE ($T_{1/2}$)

If the life of a radioactive nuclide is taken to mean the time that elapses before a the nuclei present disintegrate, then it is clear from equation [11.2] (or fro~ Fig. 11.1) that the life of any radioactive nuclide is infinite, i.e. $N = 0$ when $t = \infty$ It is not very useful, therefore, to talk about the life of a radioactive nuclide, an~ instead we refer to its half-life. **The half-life of a radioactive nuclide is the tim~ taken for half the nuclei present to disintegrate.** If the half-life is represente~ by $T_{1/2}$, then when $t = T_{1/2}$, $N = N_0/2$, and therefore by equation [11.2]

$$\frac{N_0}{2} = N_0\, e^{-\lambda T_{1/2}}$$

i.e. $\quad \frac{1}{2} = e^{-\lambda T_{1/2}}$

i.e. $\quad \log_e\left(\tfrac{1}{2}\right) = -\lambda T_{1/2}$

i.e. $\quad -0.6931 = -\lambda T_{1/2}$

i.e. $\quad \boxed{T_{1/2} = \dfrac{0.6931}{\lambda}}$

The concept of half-life is illustrated in Fig. 11.1. The reader should verify, b~ selecting any point on the curve, that the number of nuclei halves whenever increases by $T_{1/2}$. The half-life of any given nuclide is constant, in particular, does not depend on the number of nuclei present.

EXAMPLE 11.1

A sample of a radioactive material contains 10^{18} atoms. The half-life of the materi~ is 2.000 days. Calculate:

(a) the fraction remaining after 5.000 days,

(b) the activity of the sample after 5.000 days.

Solution

(a) Since $N = N_0\, e^{-\lambda t}$, the fraction, N/N_0, remaining after time t is given by

$$\frac{N}{N_0} = e^{-\lambda t}$$

Here $t = 5.000$ days and $\lambda = 0.6931/2.000 \text{ day}^{-1}$.

$$\therefore \quad \lambda t = \frac{0.6931}{2.000} \times 5.000 = 1.7328$$

$$\therefore \quad \frac{N}{N_0} = e^{-1.7328} = 0.1768$$

i.e.   Fraction remaining after 5.000 days $= 0.1768$

(There has been no need to express t in s, nor to express λ in s^{-1}. We are concerned with λt, which is a pure number and therefore any unit of time can be used for t as long as the reciprocal of the same unit is used for λ.)

(b)
$$\frac{dN}{dt} = -\lambda N$$

Here

$$N = 0.1768 \times 10^{18} \quad \text{and} \quad \lambda = \frac{0.6931}{2.000 \times 24 \times 3600} s^{-1}$$

$$\therefore \quad \frac{dN}{dt} = -\frac{0.6931 \times 0.1768 \times 10^{18}}{2.000 \times 24 \times 3600}$$

$$= -7.092 \times 10^{11} s^{-1}$$

i.e. Activity after 5.000 days $= 7.092 \times 10^{11}$ Bq

EXAMPLE 11.2

A sample of radioactive material has an activity of 9.00×10^{12} Bq. The material has a half-life of 80.0 s. How long will it take for the activity to fall to 2.00×10^{12} Bq?

Solution

Since activity (rate of decay) is proportional to the number of parent nuclei present (see section 11.4), it follows from

$$N = N_0 e^{-\lambda t}$$

that

$$A = A_0 e^{-\lambda t}$$

where A = activity at time t, and A_0 = activity at $t = 0$. Rearranging gives

$$\frac{A}{A_0} = e^{-\lambda t}$$

$$\therefore \quad \log_e \left(\frac{A}{A_0} \right) = -\lambda t$$

$$\therefore \quad \log_e \left(\frac{2.00 \times 10^{12}}{9.00 \times 10^{12}} \right) = -\frac{0.6931}{80.0} \times t$$

$$\therefore \quad -1.504 = -8.664 \times 10^{-3} t$$

$$\therefore \quad t = 174\,s$$

i.e. Time for activity to fall to 2.00×10^{12} Bq is 174 s.

QUESTIONS 11B

1. The half-life of a particular radioactive material is 10 minutes. Without using a calculator, determine what fraction of a sample of the material will decay in 30 minutes.

2. A radioactive source has a half-life of 20 s, and

an initial activity of 7.0×10^{12} Bq. Calculate it activity after 50 s have elapsed.

3. A sample of radioactive waste has a half-life c 80 years. How long will it take for its activity t fall to 20% of its current value?

11.6 PHYSICAL, BIOLOGICAL AND EFFECTIVE HALF-LIFE

Most substances that are taken into the body are subsequently removed b processes such as urination and defecation. If the substance is radioactive, it activity within the body will therefore fall more quickly than it would if it were du to radioactive decay alone, i.e. the **effective half-life** of the material will be les than its **physical (radioactive) half-life**.

The amount of substance remaining in the body often decreases exponentiall with time, and in such cases it is meaningful to define a quantity called th biological half-life.

> The **biological half-life** T_B of a material is the time taken for half the material to be removed from the body by biological processes.

If λ_B and λ_R are respectively the fractions of the radioactive nuclei removed per un time by biological processes and by radioactive decay, then the total fractio removed per unit time, λ_E, is given by

$$\lambda_E = \lambda_B + \lambda_R$$

Bearing in mind that the fraction removed per unit time is the decay constant (se section 11.4), it follows that

$$\frac{0.6931}{T_E} = \frac{0.6931}{T_B} + \frac{0.6931}{T_R}$$

$$\therefore \quad \boxed{\frac{1}{T_E} = \frac{1}{T_B} + \frac{1}{T_R}}$$

where T_E and T_R are the effective half-life and the physical (radioactive) half-lif respectively.

QUESTIONS 11C

1. A radionuclide has a half-life of 15 days and a biological half-life of 10 days. If 2.4 g of the nuclide are injected into a patient, what mass of it is there in the patient's body 18 days later?

2. The radionuclide iodine 131 has a half-life of days. A patient injected with 2.00 g of iodin 131 has only 0.125 g of it in his body 24 day later. Find (a) the effective half-life, (b) th biological half-life of iodine 131.

11.7 NATURE AND PROPERTIES OF α, β AND γ

α-particles

An α-particle consists of two protons and two neutrons, i.e. it is identical to a helium nucleus. The velocity with which an α-particle is emitted depends on the species of nucleus (i.e. the nuclide) which has produced it and is typically 6% of the velocity of light. This corresponds to a kinetic energy of 6 MeV and they are the most energetic form of 'radiation' produced by radioactive decay.

Since α-particles are charged and move relatively slowly, they produce considerable ionization ($\sim 10^5$ ion-pairs* per cm in air at atmospheric pressure). As a consequence they lose their energy over a short distance and, for example, are capable of penetrating only a single piece of paper or about 5 cm of air. They are not easily deflected and their paths through matter are largely straight.

β-particles

These are very fast electrons (up to 98% of the speed of light). In spite of their great velocities, they have less energy than α-particles on account of their much smaller mass.

Because they are less massive than α-particles, they are much more easily deflected and their paths through matter are therefore tortuous. They produce much less ionization than α-particles ($\sim 10^3$ ion-pairs per cm of air at atmospheric pressure). The most energetic are about 100 times more penetrating than α-particles.

γ-rays

γ-rays are electromagnetic radiation of very short wavelength. Wavelengths are typically in the range 10^{-10} m to 10^{-12} m, corresponding to energies of about 0.01 MeV to about 1 MeV. It is not uncommon for X-rays to have a wavelength of 10^{-11} m. These differ from γ-rays of the same wavelength only in the manner in which they are produced; γ-**rays are a result of nuclear processes, whereas X-rays originate outside the nucleus.**

In comparison with α-particles and β-particles, γ-rays produce very little ionization and therefore have much greater penetration. (90 m of air reduces the intensity of a beam of 1 MeV γ-rays by 50%.) The various mechanisms by which γ-rays produce ionization are the same as those described for X-rays in section 10.6.

Note   The intensity of a <u>monoenergetic</u> beam of γ-rays decreases exponentially with absorber thickness – equation [10.1] applies to both X-rays and γ-rays.

11.8 THE INVERSE SQUARE LAW FOR γ-RAYS

A point source of γ-rays emits in all directions about the source. It follows that the intensity of the γ-radiation decreases with distance from the source because the rays are spread over greater areas as the distance increases. This decrease in intensity is distinct from that produced by absorption.

---

* Ionization results in the release of two charged particles – an electron and a positive ion, collectively known as an **ion-pair**.

Consider a point source of γ-rays, situated in a vacuum so that there is n< absorption. The radiation spreads in all directions about the source, and therefor< when it is a distance d from the source it is spread over the surface of a sphere o radius d and area $4\pi d^2$. If E is the energy radiated per unit time by the source, then the intensity of the radiation (= energy per unit time per unit area) is given by *I* where

$$I = E/(4\pi d^2)$$

i.e. $I \propto 1/d^2$

Thus the intensity varies as the inverse square of the distance from the source.

Note   The inverse square law holds for <u>all</u> forms of radiation <u>in vacuum</u> providing th< radiation is coming from a <u>point</u> source. It can also be taken to apply to γ-rays and X-rays <u>in air</u> because these particular forms of radiation are absorbed only slightl< by air.

QUESTIONS 11D

1. A Geiger counter placed 20 cm from a point source of γ-radiation registers a count rate of $6000\,s^{-1}$. What would the count rate be 1.0 m from the source?

2. A point source of γ-radiation is moved 5 m closer to a Geiger counter and the count rat< increases from $36\,s^{-1}$ to $256\,s^{-1}$. How far from the counter was the source originally?

CONSOLIDATION

Nuclei contain protons and neutrons.

A nucleon is a proton or a neutron.

Atoms which have the same number of protons are atoms of the same **element**.

Two atoms which have the same number of protons but different numbers o< neutrons are **isotopes** of each other.

Atomic number (or **proton number**) Z of an element is the number of proton: in the nucleus of an atom of the element.

Mass number (or **nucleon number**) A of an atom is the number of nucleons (i.e protons + neutrons) in its nucleus.

Radioactive decay is spontaneous and therefore the rate of decay i< proportional to the number of parent nuclei present. This leads to

$$\frac{dN}{dt} = -\lambda N$$

where λ is a constant of proportionality called the **decay constant**.

Solving this equation gives **the exponential law of radioactive decay**

$$N = N_0\,e^{-\lambda t}$$

Because activity, A, and count-rate, R, are proportional to N, it follows that

$$A = A_0\,e^{-\lambda t} \quad \text{and} \quad R = R_0\,e^{-\lambda t}$$

Activity is the rate of decay, i.e. the number of disintegrations per second. The unit of activity is the becquerel.

One becquerel (Bq) = 1 disintegration per second.

$$\text{Activity} = -\frac{dN}{dt} = \lambda N$$

Half-life $(T_{1/2})$ is the time taken for half the nuclei present to disintegrate.

$$T_{1/2} = \frac{\log_e 2}{\lambda} = \frac{0.6931}{\lambda}$$

α-**particles** consist of two protons and two neutrons.

β-**particles** are (fast) electrons.

γ-**rays** are short-wavelength electromagnetic radiation.

The inverse square law for γ-rays. The intensity, I, at a distance, d, from a point source of γ-rays is given by

$$I \propto \frac{1}{d^2} \quad \text{i.e.} \quad I = \frac{\text{constant}}{d^2}$$

The biological half-life T_B of a material is the time taken for half the material to be removed from the body by biological processes.

QUESTIONS ON CHAPTER 11

1. A certain α-particle track in a cloud chamber has a length of 37 mm. Given that the average energy required to produce an ion-pair in air is 5.2×10^{-18} J and that α-particles in air produce on average 5.0×10^3 such pairs per mm of track, find the initial energy of the α-particle. Express your answer in MeV. (Electron charge = 1.6×10^{-19} C.)   [C, '83]

2. A radioactive source has a half-life of 20 days. Calculate the activity of the source after 70 days have elapsed if its initial activity is 10^{10} Bq.

3. A sample of iodine contains 1 atom of the radioactive isotope iodine 131 (^{131}I) for every 5×10^7 atoms of the stable isotope iodine 127. Iodine has a proton number of 52 and the radioactive isotope decays into xenon 131 (^{131}Xe) with the emission of a single negatively charged particle.
   (a) State the similarities and differences in composition of the nuclei of the two isotopes of iodine.

   (b) What particle is emitted when iodine 131 decays? Write the nuclear equation which represents the decay.
   (c) The diagram shows how the activity of a freshly prepared sample of the iodine varies with time. Use the graph to determine the decay constant of iodine 131. Give your answer in s^{-1}.

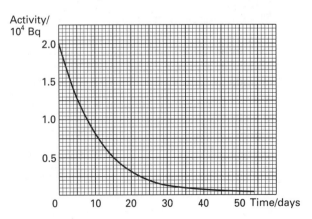

   (d) Determine the number of iodine atoms in the original sample.   [AEB, '91]

4. (a) Explain why the *physical half-life* of a radionuclide in a biological system differs from its *effective half-life* and state what is meant by the *biological half-life*. Write down an equation which relates these three quantities to each other.

    (b) The physical half-life of a radio-isotope is 20 days. Immediately after administration of the isotope to a patient the count rate at a particular site on the body surface is observed to be $2500\,s^{-1}$. Five days later, the same site gave a reading of $1000\,s^{-1}$. Find the biological half-life of the isotope.
    [N, '91]

5. The radioactive isotope $^{218}_{84}Po$ has a half-life of 3 min, emitting α-particles according to the equation

    $$^{218}_{84}Po \rightarrow \alpha + {}^{x}_{y}Pb.$$

    What are the values of x and y?
    If N atoms of $^{218}_{84}Po$ emit α-particles at the rate of $5.12 \times 10^4\,s^{-1}$, what will be the rate of emission after $\frac{1}{2}$ hour? [S, '80]

6. Radioactive sodium for medical purposes is produced by neutron bombardment of the stable isotope $^{23}_{11}Na$. What is a suitable source of neutrons for this reaction? Write an equation for this reaction.
    The sample thus produced contains a large percentage of stable sodium atoms, called carriers. In some circumstances, such as the decay of $^{131m}_{52}Te$ to $^{131}_{53}I$, the radioisotope produced is almost carrier free. Give the equation for the $^{131m}_{52}Te$ to $^{131}_{53}I$ decay. Explain why a carrier-free isotope can be obtained from this decay but not from the sodium reaction.
    The function of the lungs can be studied using a radioactive gas such as $^{133}_{54}Xe$ or $^{81m}_{36}Kr$. Using the information given below, state one advantage and one disadvantage of using Xe in preference to Kr for such investigations.

	Emission products	Half-life
$^{133}_{54}Xe$	β, γ	5.3 days
$^{81m}_{36}Kr$	γ	13 seconds

    [L, '93]

7. The radioactive isotope of iodine, ^{131}I, has a half-life of 8.0 days and is used as a tracer in medicine. Calculate:

    (a) the number of atoms of ^{131}I which must be present in a patient when she is tested to give a disintegration rate of $6.0 \times 10^5\,s^{-1}$

    (b) the number of atoms of ^{131}I which must have been present in a dose prepared 24 hours before. [N, '91]

8. In 420 days, the activity of a sample of polonium, Po, fell to one-eighth of its initial value. Calculate the half-life of polonium. Give the numerical values of a, b, c, d, e, f in the nuclear equation.

    $$^{a}_{b}Po \rightarrow {}^{c}_{d}\alpha + {}^{206}_{82}Pb + {}^{e}_{f}\gamma$$

9. A point source of γ-radiation has a half-life of 30 minutes. The initial count rate, recorded by a Geiger counter placed 2.0 m from the source, is $360\,s^{-1}$. The distance between the counter and the source is altered. After 1.5 hour the count rate recorded is $5\,s^{-1}$. What is the new distance between the counter and the source? [L, '77]

10. A cardiac pacemaker is a device which is used to ensure that a faulty heart beats at a suitable rate. In one pacemaker the required electrical energy is provided by converting the energy of radioactive plutonium-238 ($^{238}_{94}Pu$). The atoms of plutonium decay by emitting 5.5 MeV *alpha* particles. The daughter nucleus is an isotope of uranium (U). The plutonium has a decay constant of $2.4 \times 10^{-10}\,s^{-1}$.

    (a) (i) How many neutrons are in the nucleus of the radioactive plutonium?

        (ii) Write down the equation representing the decay indicating clearly the atomic (proton) number and the mass (nucleon) number of each nucleus.

    (b) (i) Define the term *half-life*.

        (ii) Calculate the half-life of plutonium-238.

        (iii) State why *alpha* particles are more suitable than either *beta* particles or *gamma* radiation for use in the power source. [AEB, '94]

12

BIOLOGICAL EFFECTS OF RADIATION AND DOSIMETRY

12.1 BIOLOGICAL EFFECTS OF IONIZING RADIATION

The cells of the body may become damaged through exposure to ionizing radiation. **The extent of the damage depends on**:

(i)   **the nature of the radiation** (see section 12.3),

(ii)   **the part of the body exposed** (rapidly dividing cells are the most susceptible),

(iii) **the dose received**, and

(iv)   **the rate at which the dose is received** (a given dose acquired over a long period allows the body time to recover and is likely to be less harmful than the same dose acquired over a shorter period).

Effects of radiation damage include skin burns, nausea, hair loss, sterility, destruction of bone marrow, changes in genetic material and a number of forms of cancer. Very high doses can cause death within a matter of days.

Radiation damages cells either by ionizing biologically important molecules such as DNA (a process known as **direct action**) or by causing chemical changes in the water content of the cell (**indirect action**). When radiation interacts with water, H and OH **free radicals** may be produced. These are electrically neutral, but each contains an unpaired electron in its outer shell and is extremely reactive as a consequence. Two OH radicals might combine to form hydrogen peroxide, H_2O_2. This is a powerful oxidizing agent, capable of damaging DNA and other biologically important molecules. Indirect action is the major source of cell damage – radiation is far more likely to interact with water than with DNA because each cell contains over 10^7 times as many water molecules.

The various effects that result from this cellular damage are classified as being either **hereditary** or **somatic**. Hereditary effects are the result of changes to cells concerned with reproduction, and are passed on to future generations. Somatic effects (e.g. skin burns) affect only the person exposed to the radiation.

The biological effects of radiation are also classified as being either stochastic (which means random) or non-stochastic.

A stochastic effect (e.g. cancer) is one for which there is no threshold dose – the probability of the effect being induced is proportional to the dose received. The severity of the effect (if it occurs) is not affected by the dose.

Non-stochastic effects (e.g. skin burns) occur only above some threshold level and increase in severity with increased dose.

Stochastic effects are due to cells mutating; non-stochastic effects are due to cells being killed or rendered unable to reproduce themselves. Hereditary effects are stochastic. Somatic effects may be either stochastic or non-stochastic. Most of the radiation risks associated with diagnostic imaging are stochastic.

Note   Cancer cells are particularly susceptible to the effects of ionizing radiation because they divide more rapidly than normal cells – this is the basis of radiation therapy.

12.2 EXPOSURE AND ABSORBED DOSE

Exposure

When a beam of ionizing radiation passes through matter, energy is absorbed from the beam. The biological damage produced by the radiation is closely related to the amount of energy absorbed. In the case of X-rays and γ-rays a measurement of the ionization that is produced provides a useful assessment of the total energy absorbed. (This is because the proportion of the energy that produces ionization, rather than heat or molecular excitation for example, is constant over a wide range of photon energies.)

For many years the ionization produced in air has been used as the basis for monitoring X-rays and γ-rays. Air is used because its (effective) atomic number is very close to that of soft tissue and it therefore has similar absorption characteristics but is a much easier medium with which to work.

Exposure is defined by

$$X = \frac{Q}{m}$$

where

X = exposure (unit = $C\,kg^{-1}$)

Q = the total charge on all the ions of any one sign ($+$ or $-$) produced in a mass m of air.

Absorbed Dose

Exposure applies only to X-rays and γ-rays in air. Absorbed dose is a more useful quantity because it applies to all forms of ionizing radiation in all materials. Furthermore, it is concerned with the energy absorbed by irradiated material rather than the ionization produced. It is defined by

$$D = \frac{E}{m}$$

where

D = absorbed dose. The unit is the **gray** (**Gy**). $1\,\text{Gy} = 1\,\text{J}\,\text{kg}^{-1}$.

E = the energy absorbed by a mass m of irradiated material.

Notes   (i)   Absorbed dose is difficult to measure directly and is usually calculated from measurements of exposure.

(ii) The unit of exposure used to be the **roentgen** (**R**).

($1\,\text{R} = 2.58 \times 10^{-4}\,\text{C}\,\text{kg}^{-1}$.)

(iii) The unit of absorbed dose used to be the **rad**. ($1\,\text{rad} = 0.01\,\text{Gy}$.)

W. K. Roentgen
discovered X-rays
in 1895

Relationship Between Exposure and Absorbed Dose

On average, the energy required to produce an ion-pair (i.e. to release one electron) in air is 34 eV. Therefore since,

$1\,\text{C}\,\text{kg}^{-1} = 1/(1.6 \times 10^{-19})$ electrons per kilogram,

$1\,\text{C}\,\text{kg}^{-1} = 34/(1.6 \times 10^{-19})\,\text{eV}\,\text{kg}^{-1}$

$\qquad\qquad = 34\,\text{J}\,\text{kg}^{-1}$

$\qquad\qquad = 34\,\text{Gy}$

Therefore

Absorbed dose in Gy $= 34 \times$ Exposure in $\text{C}\,\text{kg}^{-1}$ (in air)

For materials other than air it is necessary to take account of the fact that the energy absorbed depends on the material concerned and on the photon energies involved. It can be shown that

Absorbed dose in Gy $= f \times$ Exposure in $\text{C}\,\text{kg}^{-1}$ (in general)

where f is a factor which depends on both the nature of the absorber and the photon energy of the X- or γ-rays – see Fig. 12.1. Note that at high photon energies f ha much the same value for all three materials. This is because Compton scatter is the dominant attenuation process at these energies and Compton scatter i independent of Z. Photoelectric absorption is the dominant process at lowe energies. This is highly dependent on Z, and therefore the value of f for bon $(Z \approx 14)$ is very different from those for air and for muscle $(Z \approx 7$ in each case)

Fig. 12.1
f-factor for converting exposure to absorbed dose

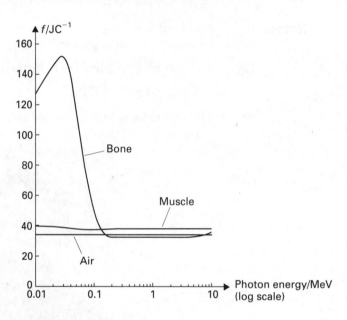

12.3 DOSE EQUIVALENT

The biological damage produced by ionizing radiation depends not only on the absorbed dose but also on the type of radiation involved. Neutrons and α-particles for example, dissipate their energy over a much shorter distance than X-rays γ-rays and β-particles and therefore can cause much more damage even when the absorbed dose is the same. A quantity called the dose equivalent has been introduced to take account of this.

The dose equivalent (H) is defined by

$$H = Q \times D$$

where

Q = a (dimensionless) quantity called the **quality factor** of the radiation (see Table 12.1)
D = the absorbed dose (Gy)

The unit of dose equivalent is the **sievert (Sv)**. ($1\,\text{Sv} = 1\,\text{J kg}^{-1}$.) Until recentl the unit was the **rem**. ($1\,\text{rem} = 0.01\,\text{Sv}$.)

Table 12.1
Quality factors of some radiation

Radiation	Quality factor (Q)
β, γ, X	1
n	5 to 20
α	20

Notes
(i) Radiations which deposit their energy in only a short distance (e.g. α-particles and fast neutrons) are said to be **high linear energy transfer (high-LET)** radiations. X-rays and γ-rays are low-LET radiations.

(ii) **Relative biological effectiveness (RBE)**, is sometimes used (incorrectly) as being equivalent to quality factor. RBE is defined by

$$\text{RBE} = \frac{\text{Absorbed dose to produce some effect with } 250\,\text{keV X-rays}}{\text{Absorbed dose to produce the same effect with the radiation concerned}}$$

(iii) α-particles are absorbed in the dead surface layers of the skin and therefore do not constitute a serious hazard unless their source is taken into the body.

(iv) Different parts of the body have different sensitivities to radiation. A quantity called the **effective dose equivalent** is used to take account of this. Each part of the body has a weighting factor assigned to it. The effective dose equivalent is the product of the dose equivalent and the relevant weighting factor. Its unit is the same as that of dose equivalent, i.e. the sievert.

12.4 INCIDENCE OF RADIATION

We are all continually exposed to low levels of ionizing radiation – the so-called **background radiation**. Some individuals will receive additional doses as a result of X-ray examinations etc. The relative contributions of the various sources of the radiation to which we are exposed are summarized in Table 12.2.

Table 12.2
Average annual effective dose equivalent for an individual in the UK (1991)

Source	Dose/μSv	Percentage	Comment
Natural sources			
Radon	1300	50.1	
Gamma-rays from the Earth and from building materials	350	13.5	
Radionuclides occurring naturally in food	300	11.6	
Cosmic rays	260	10.0	Aircrew may receive up to 10 times this amount
Artificial sources			
Medical uses	370	14.3	Excluding radiotherapy
Occupational sources	7	0.27	
Radioactive fallout	5	0.19	This has been falling since 1962 apart from a 'blip' in 1986 due to the Chernobyl accident
Radioactive discharges	0.4	0.02	
Consumer products	0.4	0.02	Luminous watches, smoke detectors etc.

(Data kindly supplied by NRPB)

Points to Note

(i) The (total) average annual dose for a resident of the UK is 2.6 mSv.

(ii) Natural sources account for 85.2% of the annual dose; artificial sources account for only 14.8%.

(iii)   The annual dose from **radon** (an α-emitting radioactive gas that seeps into the atmosphere from uranium-bearing rocks and which increases the risk of lung cancer) is, at 50.1% on average, greater than that from any other single source. Exposure to radon depends very much on geographical location (particularly high levels occur in Cornwall and Devon for example). It also depends on the extent to which the gas is allowed to accumulate inside buildings – very much higher levels are found in some buildings than others.

(iv)   Medical uses of radiation (primarily X-ray procedures) account for over 96% of the average radiation dose from artificial sources. The dose from X-rays declined in the 1980s due to improved techniques and to the introduction of ultrasound. It has recently started to rise due to the increased use of computed tomography (CT) – the dose from a typical CT scan is about 10 mSv, the average value for a conventional X-ray examination is only about 1 mSv.

(v)   The enormous (localized) doses used in therapy ($\sim$50 Sv) are not included in the data in Table 12.2.

(vi)   By law, the **maximum permitted dose level** (**MPL**) for a radiation worker is 50 mSv per year. The National Radiological Protection Board (NRPB) has recommended that this should be reduced to 20 mSv per year. Lower levels apply to the population as a whole.

(vii)   It is thought that there is no safe level of exposure, i.e. that there is some risk of harmful effects however low the dose.

(viii)   The values given in Table 12.2 are average values. Some individuals will have considerably larger doses because of where they live, their occupation, the amount of X-ray treatment they receive, etc.

CONSOLIDATION

The cells of the body may mutate, die or fail to reproduce themselves as a result of exposure to ionizing radiation.

Hereditary effects of exposure to ionizing radiation are passed on to future generations. **Somatic effects** affect only the person exposed.

Stochastic effects have no threshold dose – the probability of the effect occurring is proportional to the dose received. **Non-stochastic effects** occur only above some threshold dose and increase in severity with increased dose.

Absorbed dose (**D**) is the energy absorbed by unit mass of irradiated material. The unit is the **gray** (**Gy**).

Dose equivalent (**H**) takes account of the fact that the biological damage produced by any given absorbed dose depends on the type of radiation involved. The unit is the **sievert** (**Sv**).

$$H = Q \times D$$

where Q is the **quality factor** of the radiation.

Effective dose equivalent takes account of the fact that different parts of the body have different sensitivities to radiation. It is the product of the dose equivalent and a weighting factor. The unit is the **sievert** (**Sv**).

Most of the ionizing radiation to which an average person is exposed comes from natural sources. Medical procedures account for nearly all of the radiation dose from artificial sources.

QUESTIONS ON CHAPTER 12

1. (a) Describe the effects of ionising radiation on the cells of living matter.
   (b) Hence explain why, when assessing a radiation hazard, the following factors must be taken into consideration:
      (i) type of radiation,
      (ii) dose rate,
      (iii) total dose.   [C, '93]

2. (a) Define the quantities *exposure* and *absorbed dose* in connection with ionising radiation. Indicate the units in which **each** is measured.
   (b) Biological damage in tissue is not only dependent on absorbed dose. Name and define a quantity which describes the relative risk from a particular radiation and state its unit of measurement.   [N, '92]

3. Explain what is meant by *dose equivalent* and why it is important in radiation dosimetry.
   Calculate the energy delivered to a person of mass 70 kg by a dose equivalent to the whole body of 30 mSv (3 rem), half the dose equivalent being acquired from radiation of quality factor one, the remainder from radiation of quality factor three.   [N, '86]

4. The effects of ionizing radiations on a biological sample depend not only on the absorbed dose but also on the types of ionizing radiations employed.
   Explain the terms *relative biological effectiveness* and *dose equivalent*.
   State **two** effects which may be observed in human beings and **two** which may be observed in individual cells exposed to ionizing radiations.
   Considering alpha particles and electrons, state
   (a) which is more damaging to tissue for equal absorbed doses,
   (b) an important difference in their behaviour which is related to the amount of damage caused.   [N, '88]

5. A radioactive source emits gamma rays and its activity in MBq is N. The rate of exposure in $C\,kg^{-1}\,h^{-1}$ at a point distance D, in m, from the source is given by $\dfrac{kN}{D^2}$ where k is a constant for the particular radionuclide making up the source.

   A certain source produces an exposure rate of $500\,\mu C\,kg^{-1}\,h^{-1}$ at a point 1 m away.
   (a) At what distance must a barrier be placed if the exposure rate at the barrier is not to exceed $25\,\mu C\,kg^{-1}\,h^{-1}$?
   (b) How many half-value thicknesses of lead would be needed to reduce the exposure rate at 1 m from the source to less than $25\,\mu C\,kg^{-1}\,h^{-1}$?
   (c) If the activity of the source were doubled, what *additional* [half-value] thickness of lead would be required to keep the exposure rate below $25\,\mu C\,kg^{-1}\,h^{-1}$ at a point 1 m away?
   (d) The average energy required to create an ion-pair in air is 5.8×10^{-18} J. Find the absorbed dose-rate in air at the barrier when the exposure rate there is $25\,\mu C\,kg^{-1}\,h^{-1}$.   [N, '90]

6. In the UK in 1988 the average annual dose of radiation of natural origin was $2000\,\mu Sv$, where the sievert (Sv) is a unit of radiation dose. Explain why a person living in a region of granite rocks, or at a high altitude, receives a dose of background radiation somewhat higher than the average.
   Concern has been expressed about the effect on radiation dose of draught proofing brick-built, centrally-heated houses. Why might the dose increase in these circumstances?   [L, '94]

7. (a) Write a brief account of the effects of ionising radiations on living matter, explaining why exposure to such radiations is considered hazardous. Radiation hazards differ depending on the type of radiation involved. Explain why this is so.
   (b) Discuss the main sources of radiation dosage to people living in Britain. How may a person minimise his radiation dosage?
   (c) Much concern has been expressed recently about the presence in buildings of *radon* (a radioactive gas).
   State
      (i) why radon is regarded as particularly hazardous,
      (ii) why radon accumulates in buildings,
      (iii) why the build-up of radon varies with the geographical location of the building.   [O & C, '91]

13

DETECTION OF RADIATION

13.1 THE FILM-BADGE DOSIMETER

Photographic film is affected by exposure to all forms of ionizing radiation – the greater the exposure, the greater the blackening when the film is developed.

The film-badge dosimeter is commonly used to monitor personnel exposed to radiation. It consists of a piece of photographic film in a light–tight envelope inside a specially designed plastic case that can be attached to the person's clothing. The film is replaced after about a month and is then processed to determine the extent of the blackening.

A film badge dosimeter

The plastic case (Fig. 13.1) contains a number of different filters so that the various types of radiation to which the wearer has been exposed can be identified. For example, the plastic filters distinguish between β-particles of different energies. Low-energy X-rays blacken the film everywhere but behind the lead filter. High-energy X-rays and γ-rays produce almost uniform blackening. By comparing the blackening under the various filters with that on a set of calibration films an estimate of the dose equivalent can be made.

Fig. 13.1
The film-badge
dosimeter

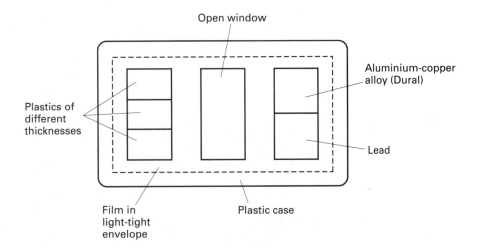

The film has two layers of emulsion – one more sensitive than the other. The sensitive (fast) layer allows low levels of exposure to be measured. High levels of exposure produce too much (saturation) blackening in this layer, and in such cases it is stripped away so that the blackening of the less sensitive (slow) layer can be used for the assessment.

Film badges are cheap, require no maintenance and provide a permanent record of exposure to a wide range of ionizing radiations. A disadvantage is the delay that occurs before the level of exposure is known.

Meaningful results can be obtained only if the badge is used sensibly. Some common mistakes include:

(i)   Leaving a lab coat with the badge attached to it hanging in a room where there is a risk of exposure.

(ii)   Laundering a lab coat without first removing the badge.

(iii)   Not always wearing the badge when there is a risk of exposure.

(iv)   Fogging the film by exposing the badge to heat or strong sunlight.

13.2 THE THERMOLUMINESCENT DOSIMETER (TLD)

Lithium fluoride (LiF) is one of a number of materials that exhibit a property known as **thermoluminescence**, and is the material most commonly used in TLDs. When ionizing radiation falls on a crystal of lithium fluoride, electrons may be raised into energy levels that are associated with defects in the crystal structure. The electrons are unable to escape from these levels at room temperature, but if the crystal is heated, they return to their normal states, emitting light in the process.

The total amount of light emitted provides an estimate of the absorbed dose. It can be measured with a photomultiplier (see section 13.6) coupled to an integrator.

Some of the factors that make lithium fluoride suitable for TLDs are listed below.

(i)   It is able to detect X-rays, γ-rays, β-particles and thermal neutrons.

(ii)   The temperature at which the light is emitted ($\sim$200 °C) is high enough to ensure stability but not so high as to cause heating problems.

(iii)   Its absorption characteristics are similar to those of soft tissue and it therefore allows direct estimates of absorbed dose to be made.

The LiF is usually incorporated into a plastic disc contained inside a light-tight case and attached to the clothing like a film badge. TLDs are both more accurate and more sensitive than film badges and have a larger usable range. Furthermore, although the initial cost is higher, a TLD 'chip' can be used many times – photographic film can be used only once.

A thermoluminescent dosimeter

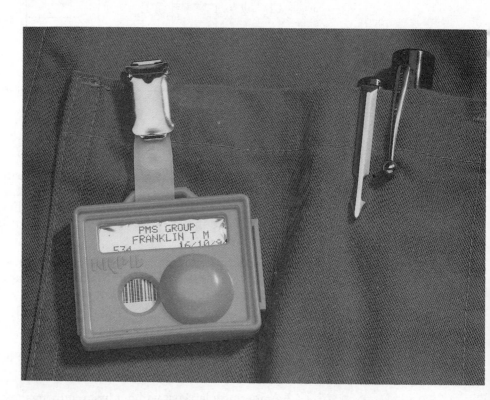

13.3 IONIZATION CHAMBERS

An ionization chamber consists of two electrodes between which there is a gas – often air at atmospheric pressure. When ionizing radiation enters the chamber it ionizes the gas. The electrons and positive ions that result are drawn towards the positive and negative electrodes respectively, causing a current to flow in the external circuit.

Fig. 13.2 illustrates the way in which this **ionization current** depends on the PD across the chamber. Between O and A the PD is not large enough to draw all the electrons and positive ions to their respective electrodes before some recombination has occurred. Between A and B the PD is large enough to prevent recombination but is not so high that it produces secondary ionization. The ionization current is said to have reached its **saturation value** (I_s). Beyond B the PD is large enough to cause secondary ionization. The PD at which an ionization chamber is operated should be such that the ionization current has its saturation value. Under such conditions:

(i)   the ionization current is independent of fluctuations in supply voltage, and

(ii)   the ionization current is proportional to the rate at which ionization is being produced in the chamber. (The reader should contrast this with the case of the Geiger–Müller tube (section 13.5), where the output is proportional to the number of ionizing particles.)

There are many different types of ionization chamber. Two that are commonly used in medicine are the thimble chamber and the condenser chamber.

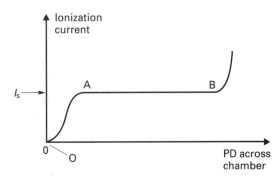

Fig. 13.2
Variation of ionization current with PD across an ionization chamber

The Thimble Ionization Chamber

The thimble ionization chamber (Fig. 13.3) measures exposure or exposure rate. The chambers used for clinical purposes consist of a small volume ($<1 \text{ cm}^3$) of air surrounded by an **air-equivalent** wall, i.e. a wall that has the same mass attenuation coefficient as air (because it has a similar effective atomic number). The wall, usually made of a plastic material impregnated with graphite, forms one electrode of the device; a centrally positioned aluminium rod, electrically isolated from the wall, is the other. Their small size makes them suitable for measuring exposure at localized points on the body, or even inside body cavities.

Fig. 13.3
Thimble ionization chamber

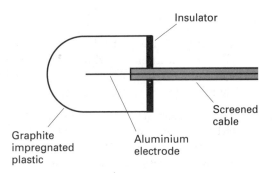

X-rays or γ-rays incident on the chamber ionize the air in it, primarily by liberating electrons from the wall – very little ionization is the result of X-rays or γ-rays being absorbed directly by the air. The PD across the chamber is such that all the electrons and ions are collected by the electrodes, i.e. there is no recombination.

When the device is being used to measure exposure (rather than exposure rate) the circuit shown in Fig. 13.4(a) is used. The charge (Q) liberated in the chamber accumulates on the capacitor. The PD (V) across the capacitor continues to build up as long as the chamber is exposed to ionizing radiation. Since $V = Q/C$, the meter reading is proportional to Q and therefore to the exposure. Furthermore, because the wall is air-equivalent, the reading is proportional to exposure over a wide range of photon energies.

Fig. 13.4
Thimble ionization
chamber circuits to
measure (a) exposure (b)
exposure rate

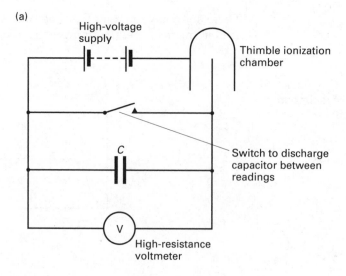

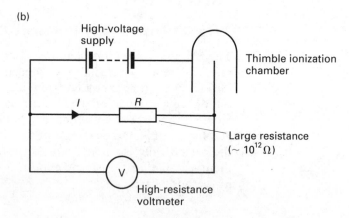

The circuit of Fig. 13.4(b) is used for measurements of exposure rate. Since $V = IR$, the meter reading is proportional to the current flowing through the resistor, and therefore to the rate at which charge is being produced in the chamber, i.e. to exposure rate.

The thickness of the chamber wall needs to be matched to the photon energy of the radiation being monitored. If the wall is too thin, there is very little X- or γ-ray absorption and very few electrons are released from the wall. Furthermore, electrons resulting from ionizing events outside the chamber will be able to enter it and produce misleading results. If the wall is too thick, most of the absorption occurs in its outer layers and the electrons that are liberated are unable to reach the air in the chamber. If a thimble chamber is to be used with higher photon energies than those for which it has been designed, the wall thickness can be increased by adding a Perspex cap.

The Condenser Ionization Chamber

This is a modified version of the thimble chamber that has the convenience of not being connected to any external circuitry whilst it is actually being exposed to radiation. Initially, the screwed plug (Fig. 13.5) is removed to allow electrical contact with the central electrode. The chamber is charged to a known voltage (V_1). The chamber is then disconnected from the charging device and the plug is replaced. When the chamber is exposed to radiation, it gradually discharges as a result of the ionization produced in it. When the exposure is complete, the plug is removed and the new, reduced voltage between the central electrode and the wall (V_2) is measured.

Fig. 13.5
The condenser ionization chamber

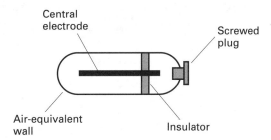

If the capacitance of the chamber is C, then (by $Q = CV$) the initial charge is CV_1 and the final charge is CV_2 and therefore

$$\text{Decrease in charge during exposure} = CV_1 - CV_2 = C(V_1 - V_2)$$

The decrease in charge is equal to the charge produced by ionization and is therefore proportional to the exposure.

13.4 THE SEMICONDUCTOR DIODE

A reverse-biased p–n junction diode* can be used as a radiation detector. When ionizing radiation enters the diode it creates electron-hole pairs by exciting electrons into the conduction band. The electrons and holes move to opposite electrodes under the influence of the applied PD and produce a current pulse in the external circuit.

These solid-state detectors are effectively solid-state ionization chambers and are potentially far more sensitive than conventional air-filled ionization chambers. There are two reasons for this.

(i)   The energy required to produce an electron-hole pair in silicon (the semiconducting material normally used) is only 3.5 eV, nearly 10 times less than the 34 eV needed to produce an electron-ion pair in a conventional ionization chamber. It follows that nearly 10 times as much charge is produced for a given amount of energy deposited by the ionizing radiation.

(ii) Solids have very much higher densities, and therefore much greater stopping powers, than gases.

13.5 THE GEIGER–MÜLLER TUBE (G–M TUBE)

A Geiger–Müller tube (Fig. 13.6) can be used to detect the presence of X-rays, γ-rays and β-particles. Tubes with very thin mica windows can also detect α-particles.

When one of these ionizing 'particles' enters the tube, either through the window or the wall, some of the argon atoms become ionized. The free electrons and

---

* Junction diodes are dealt with in most core texts – see, for example, R.Muncaster, *A-level Physics*, section 55.6.

Fig. 13.6
A Geiger–Müller tube
and circuit

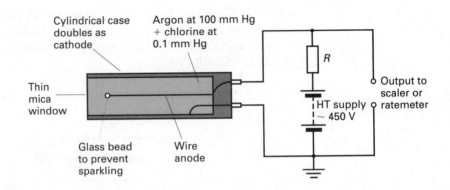

positive ions which result are accelerated towards the anode and cathode respectively by the PD across the tube. The geometry of the tube is such that the electric field near the anode is very intense, and as electrons approach the anode they gain sufficient kinetic energy to produce further ionization. The electrons released by this 'secondary' ionization produce even more ionization so that there is soon a large number of electrons moving towards the anode – the resistance of the gas is said to have broken down. The positive ions are much more massive, and move much more slowly, than the electrons, and after about 10^{-6} s there are so many positive ions near the anode that the electric field around it is cancelled out. This prevents further ionization, and the **electron avalanche** and the associated anode current, cease to exist. Thus, the effect of a single ionizing 'particle' entering the tube is to produce a relatively large current pulse. The process is called **gas amplification** and as many as 10^8 electrons can be released as a result of a single ionizing event.

The positive ions move slowly towards the cathode. Some of the ions would release electrons from the cathode surface if they were allowed to collide with it. These electrons would initiate a second avalanche, and this would give rise to a third, and so on, so that a whole series of current pulses would be produced. This would make it impossible to know whether a second ionizing 'particle' had entered the tube. In order to ensure that only one pulse is produced by each 'particle' that enters it the tube contains a **quenching agent** – chlorine. (Bromine is used in some cases.) The argon ions are neutralized as a result of collisions with chlorine molecules before they reach the cathode, and, in effect, their energy is used to dissociate the chlorine molecules rather than to release electrons from the cathode.

A resistor, R, of about 1 MΩ, is connected in series with the tube and the HT supply. The current pulse from the tube creates a voltage pulse of about 1 V across R, and this can be amplified and fed to a **scaler counter** or a **ratemeter**. A scaler registers the number of pulses it receives whilst it is switched on; a ratemeter indicates the rate at which it receives pulses and registers it in counts per second.

Immediately after a pulse has been registered there is a period of about 300 μs during which the tube is insensitive to the arrival of further ionizing 'particles'. This can be divided into two parts – the **dead time** and the **recovery time**. During the dead time the tube does not respond at all to the arrival of an ionizing 'particle'. The recovery time is the second stage of the period of insensitivity, and during this time pulses are produced but they are not large enough to be detected. The dead time is the time taken by the positive ions to move far enough away from the anode for the electric field there to return to a level which is large enough for an avalanche to start. The recovery time is the time which elapses while the argon ions are being neutralized by the quenching gas. The period of insensitivity limits the count rate to a maximum of about 1000 counts per second.

Apart from those which arrive during the period of insensitivity, almost every α-particle and β-particle that enters a Geiger–Müller tube is counted. γ-rays and X-ray photons are more likely to be detected indirectly, as a result of being absorbed by the walls of the tube and releasing electrons in the process, than by direct ionization of argon atoms. They are only weakly absorbed and only about 1% are detected i.e. **Geiger–Müller tubes have low efficiency for X- and γ-rays**.

Every 'particle' that produces any ionization at all gives rise to a voltage pulse, and is therefore detected. The size of the pulse is independent of the original amount of ionization. For example, an α-particle that produces 10^5 ion-pairs (directly) creates a pulse of the same size as a γ-ray that has produced only a single ion-pair. The output of a G–M tube is therefore proportional to the number of ionizing 'particles', rather than to the amount of ionization they produce. It provides no information about the nature of the radiation nor the energies of the individual 'particles'.

Geiger counters are very sensitive instruments and are widely used to monitor low levels of radiation.

13.6 SCINTILLATION COUNTERS

Scintillators are materials that emit tiny flashes of light (**scintillations**) when ionizing radiation is incident on them. The light is produced when atoms, which have been excited by the radiation, return to their ground states. Each interacting particle (or photon) produces just one scintillation, the energy of which is proportional to the energy lost by the particle (or photon) in passing through the scintillator.

The scintillations are detected by means of a **photomultiplier**, a device that converts each scintillation into an electrical pulse. The height of each pulse is proportional to the energy dissipated in the scintillator by the particle (or photon) that initiated it. By using a multichannel pulse-height analyser to count the number of pulses in each of a number of narrow height ranges, it is therefore possible to obtain the spectrum of the incident radiation.

The most widely used scintillators are crystals of sodium iodide to which about 0.5% thallium iodide has been added. The thallium is known as an **activator** and its presence greatly increases the light output of the crystal. Some of the of reasons for the popularity of **sodium iodide** are listed below.

(i) It converts the energy of the incident radiation into light with high efficiency.

(ii) The duration of each scintillation is short and therefore high count rates can be achieved.

(iii) It does not absorb any of the light it produces.

Sodium iodide is hygroscopic and is therefore kept in sealed containers. These are normally aluminium cans with a glass window on the side which is attached to the photomultiplier.

Sodium iodide is normally used for the detection of γ-rays and high-energy X-rays. The relatively high atomic number of iodine ($Z = 53$) means that the dominant attenuation mechanism is photoelectric absorption and therefore the whole of the energy of the incident photon is utilised (see section 10.6). Other scintillators in common use include zinc sulphide (for α) and lithium iodide (for thermal neutrons). Anthracene, an organic material, is used for β-particles and fast neutrons.

The Photomultiplier

Refer to Fig. 13.7. Light from the scintillator falls on the photocathode of the photomultiplier and ejects electrons from it. The electrons are accelerated towards a series of electrodes known as **dynodes**. On impact with a dynode each electron ejects about 4 more electrons by the process of **secondary emission**. This happens at each of 10 dynodes, producing an overall amplification (or **multiplication factor**) of 4^{10}.

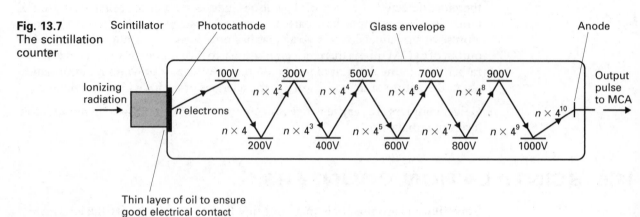

Fig. 13.7
The scintillation counter

The sensitivity of the instrument can be improved by reducing the amount of noise present. This is achieved by:

(i)   cooling it to reduce thermionic emission from the photocathode and dynodes,

(ii) shielding it from stray radiation.

The scintillation counter has a number of advantages over other detectors.

(i)   It is extremely sensitive to X-rays and γ-rays.

(ii) It allows the energy spectrum of the incident radiation to be obtained, and this can often be used to identify the source nuclide.

(iii) The various scintillators that are available allow it to measure all forms of ionizing radiation.

QUESTIONS ON CHAPTER 13

1. Describe a film badge of the type used for monitoring the radiation exposure of personnel. Explain how it is possible to distinguish
   (a) between high and low energy X (or gamma) radiation falling on the badge and
   (b) between X (or gamma) radiation and beta-radiation falling on the badge.
   What do you think limits the minimum and the maximum doses which can be measured by a film badge?   [N, '90]

2. (a) A film badge used for personal radiation monitoring contains various filters through which radiation must pass before reaching the film. Explain how this helps in making an estimate of the dose equivalent received by the wearer of the badge.

   (b) Describe the principle of operation of the thermoluminescent dosimeter.   [N, '86]

3. **(a)** Draw a labelled diagram of a thimble ionisation chamber and describe briefly its mode of operation. Explain, with the aid of a circuit diagram, how the device can be used to measure exposure rate.

    (b) A radiation worker carries a thimble ionisation chamber to monitor exposure. The device has a working volume of $1.0 \times 10^{-5} \, m^3$ and contains air at S.T.P. During a normal 40-hour working week the measured steady current through the chamber is $2.0 \times 10^{-14} \, A$. Calculate the exposure.
    (Density of air at STP $= 1.29 \, kg \, m^{-3}$, $e = 1.6 \times 10^{-19} \, C$)   [N, '92]

4. Describe the physical principles underlying the action of a scintillation detector and photomultiplier, i.e. a scintillation counter.
    Why is a thallium-doped sodium iodide crystal [NaI (T1)] often used as the scintillator material for the detection of γ-rays?   [L, '88]

5. **(a)** Gamma radiation may be detected using a scintillation counter which consists of the scintillator coupled to a photomultiplier tube.
    (i) Draw a labelled diagram of a scintillation counter including relevant details of the photomultiplier tube.
    (ii) What is meant by the *multiplication factor* of the photomultiplier tube?
    (iii) Explain how the effects of background radiation are minimised.

    (b) A radioactive source is placed in front of a scintillation counter which then produces an anode current of $1.0 \times 10^{-8} \, A$ in the photomultiplier tube. The tube has a multiplication factor of 1.0×10^6. Calculate
    (i) the number of electrons per second flowing in the anode circuit,
    (ii) the number of photoelectrons per second produced at the photocathode.
    ($e = 1.6 \times 10^{-19} \, C$)   [N, '94]

6. The figure below shows a simplified diagram of a typical scintillation counter.
    (a) (i) Explain the function of the scintillator.
    (ii) Suggest why a scintillator used for detecting γ-rays may not detect α-particles.
    (iii) Describe the function of the photocathode.

    (b) The scintillation counter in the figure is used to detect β-particles. The anode current is $6.4 \times 10^{-13} \, A$. Each electron incident on a dynode results in four electrons moving on to the next dynode.
    (i) Calculate the number of electrons arriving at the anode in one second.
    (ii) There are eight dynodes in the scintillation counter.
    Calculate the number of electrons incident on the first dynode D_1 in one second.
    (iii) The rate of arrival of β-particles at the scintillator is $4000 \, s^{-1}$.
    Calculate the ratio

$$\frac{\text{rate of production of electrons by the photocathode}}{\text{rate of arrival of } \beta\text{-particles at the scintillator}}$$

[C, '95]

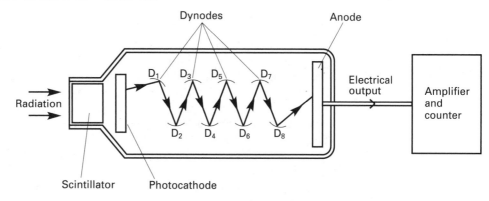

14

MEDICAL USES OF RADIOISOTOPES

14.1 INTRODUCTION

Radioisotopes are used in diagnosis, both as tracers and for imaging purposes, and in therapy. Prior to the development of the lithium iodide battery, they were also commonly used to power artificial heart pacemakers. (α-emitters with half-lives of fifty to a hundred years were used in order to minimize shielding difficulties and to avoid the need for replacement.)

14.2 RADIONUCLIDE IMAGING

If a radioisotope with an affinity for a particular organ is administered to a patient, it collects in the organ concerned and a **rectilinear scanner** or **gamma camera** can be used to produce an image of the organ by measuring the γ-ray emission from the various parts of it. (Note γ-emitters are used rather than α- or β-emitters because α- and β-particles would be absorbed within the body and therefore would not reach the detector.)

The Rectilinear Scanner

A radiation detector (e.g. a scintillation counter) is scanned back and forth across the patient in a zigzag fashion. The detector is fitted with a lead collimator so that it receives only that radiation that has come from a small region directly in front of it. The movement of the detector is replicated by that of a small lamp which traces the same zigzag pattern across a sheet of photographic film. The brightness of the lamp, and therefore the exposure of the film, is proportional to the output of the detector.

A major disadvantage of the device is the length of time it takes to build up the image. A whole-body scan can take 40 minutes or more – a long time for a patient to keep still.

Rectilinear scanners have now been superseded, for most applications, by the gamma camera.

The Gamma Camera

This consists of a large diameter (40 cm) crystal of sodium iodide about 1 cm in thickness with an array of up to 75 photomultiplier tubes mounted just above it. Immediately beneath the crystal is a collimator – a circular slab of lead pierced with thousands of narrow channels whose axes are at right angles to the crystal face (Fig. 14.1). This arrangement ensures that all the γ-rays that reach any particular part of the crystal have come from a point directly below that part. It follows that the amount of light produced at any particular point in the crystal corresponds to the γ-ray activity in that part of the patient which is directly below that point.

Fig. 14.1
a) The gamma camera
b) Arrangement of photomultiplier tubes

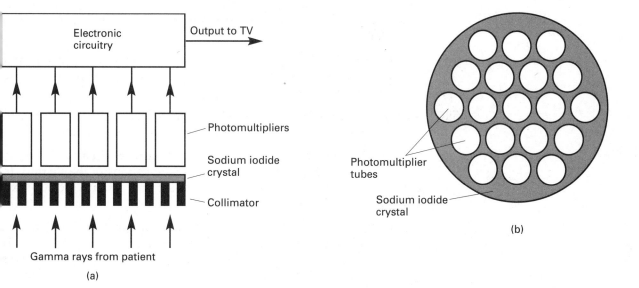

(a)

(b)

Image produced using a gamma camera

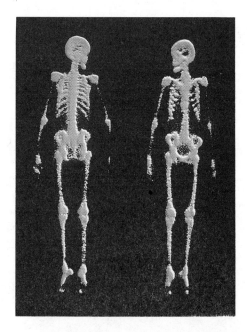

Every photomultiplier tube picks up some light from each scintillation. The various tubes are at different distances from the scintillation and therefore their outputs differ in size. The strengths of the signals are compared electronically enabling the position of the scintillation to be deduced and displayed on a TV screen. The final image is composed of up to 10^6 tiny dots, and its brightness at any point represents the γ-ray activity at the corresponding point in the patient.

The gamma camera, unlike the rectilinear scanner, is sensitive to the whole of its field of view at all times and therefore can be used to monitor rapidly changing processes.

14.3 FACTORS AFFECTING THE CHOICE OF RADIONUCLIDES USED FOR DIAGNOSIS

(i)   Ideally, the radionuclide should emit γ-rays only. α- and β-particles are absorbed within the body and therefore increase the patient's absorbed dose whilst serving no useful purpose.

(ii) The energy of the γ-rays should not be so low that they are significantly attenuated by body tissue, yet should not be so high that they are not absorbed by the detector.

(iii) The effective half-life (see section 11.6) should be as short as is practicable so that the patient does not receive an excessive dose of radiation.

(iv) The radioisotope often needs to have an affinity for a particular organ. Iodine 123 and iodine 131 are used in investigations of the thyroid, for example, because the thyroid takes up iodine to a very much greater extent than the surrounding tissue.

(v)   Many organs have an affinity for a particular compound rather than for a particular element. In such cases, the radioisotope must be one that can be chemically incorporated into the compound concerned. Compounds that have been **labelled** in this way are known as **radiopharmaceuticals**.

14.4 TECHNETIUM 99^m (^{99}Tc^m)

The most widely used radionuclide for diagnostic purposes is technetium 99^m. It is metastable (hence the 'm') and decays to technetium 99 by emitting only γ-rays.

$$^{99m}_{43}\text{Tc} \rightarrow {}^{99}_{43}\text{Tc} + \gamma$$

It is usually used in the form of a radiopharmaceutical. Albumen labelled with ^{99}Tc^m, for example, is used to study blood flow in the lungs.

There are several reasons for the widespread use of ^{99}Tc^m.

(i)   Its (physical) half-life is 6 hours. This is long enough for the investigation to be carried out but is not so long that the patient continues to be exposed to high levels of radiation after it is completed.

(ii) No α- or β-particles are emitted. These would be absorbed within the patient, contributing nothing to the image but increasing the dose received by the patient.

(iii)   The γ-rays have an energy of 140 keV. This is within the range of γ-ray energies for which the gamma camera (section 14.2) is most sensitive.

(iv)   It can be incorporated into a wide variety of radiopharmaceuticals.

(v)   It is relatively cheap.

14.5 THE MOLYBDENUM–TECHNETIUM GENERATOR

The molybdenum–technetium generator makes it possible to produce $^{99}\mathrm{Tc^m}$ relatively cheaply <u>within the hospital department where it is to be used</u>. This is an important factor because the six-hour half-life of $^{99}\mathrm{Tc^m}$ limits the time available for transportation from its production site.

The generator contains $^{99}\mathrm{Mo}$ in the form of ammonium molybdenate adsorbed on an alumina column inside a small glass tube. The $^{99}\mathrm{Mo}$ decays by β-emission with a half-life of 67 hours to produce $^{99}\mathrm{Tc^m}$ in the form of the pertechnate ion $(\mathrm{TcO_4})^-$.

$$^{99}_{42}\mathrm{Mo} \rightarrow {}^{99\mathrm{m}}_{43}\mathrm{Tc} + {}^{0}_{-1}\beta + \gamma + \bar{\nu}$$
(antineutrino)

When the $^{99}\mathrm{Tc^m}$ is required, salt (sodium chloride) solution is passed through the column. Chloride ions exchange with pertechnate ions to produce sodium pertechnate in solution with the sodium chloride. The process is known as **elution**. The solution passes into a collecting vial which is then removed from the generator so that the solution can be diluted and portioned into individual patient doses.

Very little $^{99}\mathrm{Tc^m}$ remains in the generator immediately after elution. It can be drawn off on a <u>daily</u> basis, though, because this allows sufficient time for its equilibrium concentration to be re-established (see Fig. 14.2).

The generator has to be replaced after about a week because by then the $^{99}\mathrm{Mo}$ concentration is too low to produce useful amounts of $^{99}\mathrm{Tc^m}$.

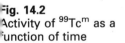

Fig. 14.2
Activity of $^{99}\mathrm{Tc^m}$ as a function of time

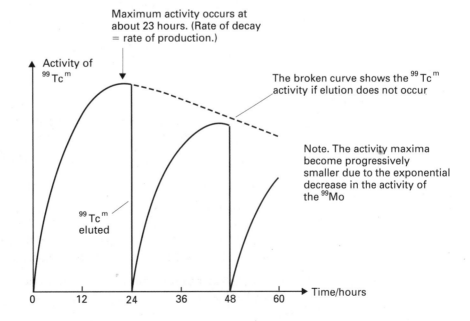

Maximum activity occurs at about 23 hours. (Rate of decay = rate of production.)

The broken curve shows the $^{99}\mathrm{Tc^m}$ activity if elution does not occur

Note. The activity maxima become progressively smaller due to the exponential decrease in the activity of the $^{99}\mathrm{Mo}$

Activity of $^{99}\mathrm{Tc^m}$

$^{99}\mathrm{Tc^m}$ eluted

Time/hours

14.6 IODINE 131 (^{131}I)

Iodine 131 is used in both diagnosis and therapy. It has a half-life of 8 days and emits β-particles accompanied by γ-rays.

$$^{131}_{53}\text{I} \rightarrow {}^{131}_{54}\text{Xe} + {}^{0}_{-1}\beta + \gamma + \bar{\nu}$$

It is produced by bombarding tellurium 130 (^{130}Te) with thermal (slow) neutrons in a fission reactor. The reaction is

$$^{130}_{52}\text{Te} + {}^{1}_{0}\text{n} \rightarrow {}^{131}_{52}\text{Te} + \gamma$$

$$^{131}_{52}\text{Te} \rightarrow {}^{131}_{53}\text{I} + {}^{0}_{-1}\beta$$

The iodine can be separated from the tellurium by standard chemical means.

14.7 TRACER STUDIES

Radioactive tracers can provide information concerning the activities of many of the organs of the body and the volumes and flow rates of various body fluids. Examples are the uptake of iodine by the thyroid and the measurement of blood volume.

Thyroid Uptake of Iodine

The thyroid gland in the neck uses iodine to manufacture hormones. The rate at which the thyroid absorbs iodine from the blood is a measure of the rate at which it is producing these hormones, and therefore of whether it is normal, underactive or overactive.

The patient is given a dilute solution of sodium iodide containing the radioactive isotope, iodine 131. An equal volume of the solution is placed in a model neck (called a **phantom**) which simulates the attenuation produced by the patient's own neck. A scintillation counter is used to compare the count rate close to the patient's thyroid with that at the same distance from the sample in the phantom. Readings are taken at various times over a period of (typically) 24 hours. The measurements are used to determine the uptake ratio, given by

$$\text{Uptake ratio} = \frac{\text{Corrected count rate from thyroid}}{\text{Corrected count rate from sample in phantom}}$$

Notes (i) The scintillation counter is surrounded by a lead collimator to shield it from radiation coming from parts of the body other than the thyroid (due to iodine still in the bloodstream, for example).

(ii) In general

Corrected count rate = Actual count rate − Background count rate

The background count rate is the count rate due to any background radiation that is present. In this case it is due to radiation that has not come from either the thyroid or the sample in the phantom. It is measured by placing a lead screen in front of the counter in each case.

(iii) The uptake ratio of a normally active thyroid 24 hours after the iodine is ingested is in the range 0.3 to 0.5.

Measurement of Body Fluids (Dilution Analysis)

This can be achieved by allowing a radioactive tracer to mix with the fluid concerned and then comparing the activity of a sample of the fluid with that of a reference sample. As an example of the method, see Example 14.1.

EXAMPLE 14.1

8.0 cm^3 of albumen labelled with iodine 131 is injected into a patient's bloodstream. When this has had time to mix thoroughly, a 5.0 cm^3 sample of blood is drawn off and is found to have an activity of 120 Bq. Another 8.0 cm^3 sample of the albumen is diluted to 5000 cm^3 with water to provide a reference sample. 5.0 cm^3 of this solution is subsequently found to have an activity of 132 Bq. Calculate the volume of the patient's blood.

Solution

Let A = activity of 8.0 cm^3 of albumen and V = patient's blood volume. It follows that

$$120 = \frac{5.0A}{V} \qquad\qquad [14.1]$$

and

$$132 = \frac{5.0A}{5000} \qquad\qquad [14.2]$$

Eliminating A between equations [14.1] and [14.2] gives $V = 5.5 \times 10^3$ cm^3.

Notes (i) We have assumed that the albumen is uniformly distributed throughout the blood, and that the volume of the albumen (8.0 cm^3) is negligible in comparison with the volume of the blood.

(ii) The albumen used as a reference was diluted. This is normal practice and is done so that the two readings can be made on the same range of the instrument, thus improving the accuracy of the final result.

14.8 RADIOISOTOPES IN THERAPY

Beams of high-energy γ-rays (usually from a cobalt 60 source) are used as an alternative to X-rays in the treatment of some forms of cancer.

There are some situations where external beams of radiation would cause too much damage to the healthy tissue surrounding the cancer cells. In such cases the cancer can be attacked from inside the body by using an implant (e.g. iridium 192) or by using a radioisotope that is taken up selectively by the organ being treated (e.g. iodine 131 for treatment of the thyroid). β-emitters are often used in these situations because they deposit their energy close to the site being treated and therefore do little damage to the surrounding tissue.

Cobalt 60 (^{60}Co)

There are a number of reasons for using ^{60}Co as an external source of γ-rays.

(i)   It produces γ-rays of sufficiently high energy (1.17 and 1.33 MeV).

(ii)   Its half-life (5.3 years) is long enough for the source not to have to be replaced too frequently but not so long that a large mass would be needed to produce the activity required ($\sim 10^{14}$ Bq).

(iii)   It is easily produced by neutron bombardment of the common isotope of cobalt (cobalt 59) in a nuclear reactor.

The cobalt source is housed inside a lead-walled container called a **head**. When it is to be used, the source is brought into line with an opening in the head by means of an electric motor. γ-rays emerge from the opening and pass through a collimator before reaching the patient.

The equipment is much simpler than that used to produce X-rays and involves none of the problems associated with high voltages. However, whereas an X-ray source can be switched off when it is not being used, a radioactive source cannot be and is therefore an ever present hazard.

Iridium 192 (^{192}Ir) Implant

Iridium 192 implants are commonly used to treat breast cancer and some cancers of the mouth. Iridium 192 emits β-particles and low-energy γ-rays. It therefore deposits most of its energy close to the site of the cancer and so does little damage to the healthy tissue beyond it. (Half-life = 74 days.)

Iodine 131 (^{131}I)

Iodine 131, which emits both β-particles and γ-rays, concentrates in the thyroid. Moderate doses ($\sim 10^8$ Bq) can be used to treat an overactive thyroid by killing some of the (healthy) cells in it and so reducing its activity. Higher doses ($\sim 10^9$ Bq) are used to treat cancer of the thyroid. (Half-life = 8 days.)

CONSOLIDATION

Radionuclides used for diagnosis should emit γ-rays only. α- and β-particles are absorbed within the body and therefore increase the patient's absorbed dose whilst serving no useful purpose.

Radionuclides used <u>inside the body</u> for therapy are usually β-emitters. They deposit their energy close to the site being treated and therefore do little damage to the surrounding tissue.

QUESTIONS ON CHAPTER 14

1. The diagram below shows a simple gamma camera.

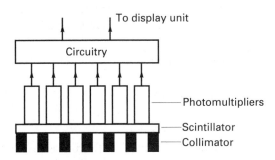

(a) Explain the purpose of the collimator.
(b) What would be a suitable substance for the scintillator?
(c) What is the function of the photomultipliers and how do they fulfil this? [L, '94]

2. (a) State **three** factors which are important in choosing a suitable radionuclide for nuclear imaging of particular parts of the body. Name **one** radionuclide in common use for this purpose.
(b) Draw a labelled sketch of the main component parts of a gamma camera. Explain how it can be used with a radionuclide to obtain the desired image of a particular part of the body.
(c) Explain how information obtained from such an image differs from that obtained from a diagnostic X-ray image. [N, '87]

3. (a) What is the principle of the use of radioactive isotopes as tracers in diagnosis? Explain what properties it is desirable for a suitable tracer isotope to have.
(b) Name **two** radioactive isotopes which are commonly used for tracer studies on several different organs or parts of the body. For *each* isotope
(i) state its approximate half-life, the type(s) of radiation it emits and how this is monitored in a medical application,
(ii) describe how it is used in one particular medical study. [O & C, '93]

4. ^{123}I is a gamma-ray emitter which can be used as a tracer in the human body. Two identical samples of ^{123}I are prepared for use in a thyroid uptake test. Explain how these are used to measure thyroid function. [N, '94]

5. A nuclear medicine investigation requires samples of the metastable radio-isotope of technetium $^{99m}_{43}$Tc, which is produced from the decay of the molybdenum isotope, $^{99}_{42}$Mo, and is eluted from the 'technetium-generator' when needed. The activities of these radioisotopes inside the generator before and after various elutions are shown in the diagram.

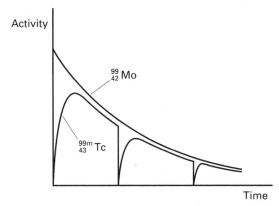

(a) Explain the shapes of the two curves.
(b) Write down the equation representing the decay of $^{99}_{42}$Mo.
(c) Write down the equation representing the decay of $^{99m}_{43}$Tc.
(d) Why is $^{99m}_{43}$Tc an 'ideal' radioisotope for diagnostic use? [L, '87]

6. (a) Describe, with the aid of a diagram, the structure and mode of action of a scintillation counter for the detection of gamma rays. State **two** advantages which such a counter has over a Geiger counter in medical applications.
(b) In a measurement of blood–plasma volume, human serum–albumen solution labelled with radioactive $^{135}_{53}$I was injected into a patient's vein. The injected volume was small compared with the plasma volume and its activity was 92 kBq. An equal volume of the serum–albumen solution was diluted with water to a volume of 3 litres and kept as a standard. 10 cm^3 of this standard gave a corrected count rate of 268 s^{-1}. After 10 minutes a blood sample was withdrawn from the patient and centrifuged to extract 10 cm^3 of plasma. This produced a corrected count rate of 252 s^{-1}. Calculate the plasma volume in the patient's body, assuming that all the injected material remained within the plasma and that the decay of $^{135}_{53}$I can be neglected.
1 litre = 1000 cm^3 [N, '89]

7. A radioactive isotope of iodine is widely used in thyroid investigations. It is produced artificially by irradiating tellurium 130 with neutrons, the resulting tellurium 131 then decaying to form the wanted iodine. (Iodine, I, and tellurium, Te, have proton numbers (atomic numbers) 53 and 52 respectively.)

(a) Write down two nuclear equations describing this process.

(b) Why is the iodine produced above suitable for thyroid studies? [L, '91]

8. In a thyroid uptake test, two identical samples of sodium iodide containing radioactive ^{131}I are prepared. One sample is given to the patient by mouth, the other is placed in a model neck at a position corresponding to the thyroid gland. After 24 hours the radioactivity due to ^{131}I (whose decay produces both β particles and γ radiation) is measured in the patient and in the model by means of a scintillation counter fitted with a collimator. In each case, count rates are observed at the position of the thyroid gland (i) with and (ii) without a lead screen in front of the counter.

The ratio

corrected count rate from the patient's thyroid gland / corrected count rate from the model neck

is compared with that determined for normal healthy subjects.

(a) What is meant by the *corrected count-rate* and why is the correction necessary?

(b) What purpose is served by the collimator? Suggest a suitable material for its manufacture.

(c) Why is it unnecessary to know the half-life of ^{131}I for this determination?

(d) Give **two** reasons why a scintillation counter is chosen rather than a Geiger counter.

(e) Name **two** mechanisms by which the γ rays (energy 0.36 MeV, 5.8×10^{-14} J) are absorbed in the lead.

(f) Will the β particles contribute significantly to the observed count rate
(i) with the lead in position;

(ii) without the lead?
Give a reason for each of your answers.

(g) Explain why, even though the measurement would be easier if a larger count-rate were achieved by use of samples of greater activity, such samples are not used.

(h) Give the reason why the patient will contain less than half the initial activity of ^{131}I after 8 days, which is the radioactive half-life of the isotope. [N, '84]

9. (a) Give **two** examples of radioisotopes which have been used for diagnostic purposes in medicine. State how **one** of these radioisotopes may be produced.

(b) Two samples, each of volume 10^{-6} m^3, contained tritiated water, the activity of each sample being 4.00 MBq (108 μCi). One sample was injected into the bloodstream of a patient and after a suitable period of time 4.00×10^{-6} m^3 of blood was withdrawn. The corrected count rate produced by this blood in a liquid scintillation counter was 207 counts per second. The other sample was diluted 10 000 times with ordinary water and 4.00×10^{-6} m^3 of the diluted liquid produced 745 counts per second, after correction, in the same scintillation counter.
Calculate the volume of water in the patient's body. The half life of tritium may be assumed to be very long compared with the duration of the experiment.

(c) Explain, for the measurements in (b),
(i) the nature of the correction applied to the count rates,
(ii) why the volume of fluid injected into the bloodstream needed to be small,
(iii) why time was allowed to elapse before withdrawing the blood for measurement,
(iv) why, apart from convenience of calculation, the dilution factor was chosen to be 10 000. [N, '81]

ANSWERS

ANSWERS TO
END OF CHAP-
TER QUES-
TIONS

Chapter 1
1. (a) $7.6 \times 10^2\,$N (b) 2.0
2. (a) $R = 200\,$N, $C = 8.0 \times 10^2$,
   $T = 6.0 \times 10^2\,$N
4. (b) (i) About $200\,$N
   (ii) About $200\,$W

Chapter 2
1. (c) (i) $47\,$kJ (ii) $20\,$g
2. (c) (i) $2.9 \times 10^2\,$W
   (ii) $5.0 \times 10^2\,$W
3. (b) (i) $1.2 \times 10^2\,$J min^{-1}
   (ii) $6.1 \times 10^2\,$J min^{-1}
4. (b) (i) About $8\,$W
5. (c) $1.1\,$W
6. (b) (ii) $197\,$g
7. (c) (ii) $29\,$m^{-1}, $25\,$m^{-1}
   (d) (ii) $5\,$MJ (iii) $0.32\,$kg

Chapter 3
1. (b) (i) $-2.0\,$D, $0.28\,$m
2. $200\,$cm (diverging), $50\,$cm (converging)
3. $4\,$D, $-1\,$D, $176\,$mm
4. (b) (iii) $2.7\,$D
5. (a) (i) $3.5\,$D
   (ii) $28.6\,$cm, $3.6\,$cm
6. (b) (ii) $40\,$cm
   (iii) $15\,$cm to $40\,$cm
7. (b) (iii) $-2\,$D (iv) $0.33\,$m
8. (b) $0.15\,$mm

Chapter 4
1. (a) $1\,$W m^{-2} (b) 0.6
2. (b) (ii) 1.1
3. (b) $0.12\,$W
4. (b) $5.0\,$mW m^{-2}, $3\,$dB
5. (b) $3.4\,$kHz
   (d) (i) $1.6 \times 10^{-9}\,$W m^{-2}
6. (b) (ii) $5.0 \times 10^{-9}\,$W m^{-2}
7. (b) (i) $1.3\,$km
   (ii) $1.0 \times 10^{-2}\,$W m^{-2}
   (iii) $2.2 \times 10^5\,$W
8. (b) (iv) $67\,$dB
9. (b) (i) $10^{-2}\,$W m^{-2}
   (ii) $40\,$dB
   (c) $1.5 \times 10^{-15}\,$W
10. (a) (i) 63
11. (b) $78\,$dB

Chapter 5
2. (a) $4.1\,$A m^{-2} (b) $2.1 \times 10^{-11}\,$A
3. (b) $0.17\,$MΩ

Chapter 6
1. (b) (i) $27\,$kPa (ii) $7.8\,$kPa
   (iii) $1.30 \times 10^4\,$Pa
2. (a) 10 (b) $27\,$kPa, $6.4\,$kPa
3. (c) (i) $8.6\,$kPa (ii) 2.7

Chapter 7
2. (b) (v) $26.8°$
3. (b) (i) $0.2\,$mm

Chapter 8
1. (a) $5.2\,$MHz
2. (c) $2.1\,$kHz
8. $1.5\,$cm s^{-1}
9. $26\,$cm s^{-1}

Chapter 10
3. (a) $0.2\,$MW m^{-2}
   (b) $12.5\,$kW m^{-2}
4. (b) (ii) $32\,$m^{-1} (iii) $14\,$cm
6. (a) (iii) $50\,\mu$W m^{-2}
8. 0.63
11. (d) $1\,$mm
16. (a) $5.0 \times 10^2\,$W m^{-2}
18. (c) $7.5\,$mm

Chapter 11
1. $6.0\,$MeV
2. $8.8 \times 10^8\,$Bq
3. (c) $1.0 \times 10^{-6}\,$s^{-1} (d) 1.0×10^{18}
4. (b) $4.7\,$days
5. $50\,$s^{-1}
7. (a) 6.0×10^{11} (b) 6.5×10^{11}
8. $140\,$days; $a = 210$, $b = 84$, $c = 4$,
   $d = 2$, $e = 0$, $f = 0$
9. $6.0\,$m
10. (a) (i) 144 (b) (ii) $2.9 \times 10^9\,$s
    ($92\,$years)

Chapter 12
3. $1.4\,$J
5. (a) $4.5\,$m (b) 5 (c) 1
   (d) $9.1\,$Gy h^{-1}

Chapter 13
3. (b) $2.2 \times 10^{-4}\,$C kg^{-1}
5. (b) (i) $6.3 \times 10^{10}\,$s^{-1}
   (ii) $6.3 \times 10^4\,$s^{-1}
6. (b) (i) 4.0×10^6 (ii) 61
   (iii) 0.015

Chapter 14
6. (b) $3.2\,$litres
9. (b) $0.036\,$m^3

ANSWERS TO
QUESTIONS
1A TO 11D

1A 1. $6.0 \times 10^2\,$N
1B 1. $3.5\,$m s^{-1}
1B 2. (a) $7.2 \times 10^2\,$J
   (b) $3.5\,$m s^{-1}
1B 3. $72\,$W
1B 4. $4.8 \times 10^2\,$W
2A 1. $154\,$kJ h^{-1} m^{-2}
2B 1. $3.2 \times 10^2\,$W
3A 1. (c) $28\,$cm (d) $3.6\,$D
   (e) $28\,$cm
3A 2. $50.5\,$D
3A 3. $67\,$cm
3A 4. (a) $400\,$cm diverging
   (b) $44\,$cm to infinity
3A 5. (a) $43\,$cm converging
   (b) $40\,$cm
4A 1. (a) $80\,$dB (b) $25\,$dB
   (c) $8.5\,$dB
4A 2. (a) $0.10\,$W m^{-2}
   (b) $6.3 \times 10^{-6}\,$W m^{-2}
4A 3. $20\,$dB
4A 4. $80.4\,$dB
4A 5. $2.0 \times 10^{-6}\,$W
6A 1. $1.03 \times 10^5\,$Pa
6A 2. $120\,$mmHg
6B 1. $24\,$mV
8A 1. (a) $4.5\,$mm
   (b) $0.90\,$mm
   (c) $0.45\,$mm
10A 1. $59\,$kV
10A 2. $4.1 \times 10^{-11}\,$m
10A 3. $40\,$mA
10A 4. (a) $87\,$kV (b) $72\,$kV
   (c) $1.4 \times 10^{-11}\,$m
10B 1. (a) Higher intensity with
   steady PD.
   (b) $100\,$keV in each case.
10C 1. (a) $1.0 \times 10^2\,$kW m^{-2}
   (b) $25\,$kW m^{-2}
10C 2. (a) $0.29\,$mm^{-1}
   (b) $2.8 \times 10^2\,$kW m^{-2}
10D 1. (a) 2:1 (b) 4:1
11A 1. (a) 26 (b) 26
   (c) 30 (d) 56
   (e) 26 (f) 26
   (g) 30
11B 1. $\frac{7}{8}$
11B 2. $1.2 \times 10^{12}\,$Bq
11B 3. $186\,$years
11C 1. $0.3\,$g
11C 2. (a) $6\,$days (b) $24\,$days
11D 1. $240\,$s^{-1}
11D 2. $8\,$m

137

INDEX